Springer Theses

Recognizing Outstanding Ph.D. Research

Aims and Scope

The series "Springer Theses" brings together a selection of the very best Ph.D. theses from around the world and across the physical sciences. Nominated and endorsed by two recognized specialists, each published volume has been selected for its scientific excellence and the high impact of its contents for the pertinent field of research. For greater accessibility to non-specialists, the published versions include an extended introduction, as well as a foreword by the student's supervisor explaining the special relevance of the work for the field. As a whole, the series will provide a valuable resource both for newcomers to the research fields described, and for other scientists seeking detailed background information on special questions. Finally, it provides an accredited documentation of the valuable contributions made by today's younger generation of scientists.

Theses may be nominated for publication in this series by heads of department at internationally leading universities or institutes and should fulfill all of the following criteria

- They must be written in good English.
- The topic should fall within the confines of Chemistry, Physics, Earth Sciences, Engineering and related interdisciplinary fields such as Materials, Nanoscience, Chemical Engineering, Complex Systems and Biophysics.
- The work reported in the thesis must represent a significant scientific advance.
- If the thesis includes previously published material, permission to reproduce this must be gained from the respective copyright holder (a maximum 30% of the thesis should be a verbatim reproduction from the author's previous publications).
- They must have been examined and passed during the 12 months prior to nomination.
- Each thesis should include a foreword by the supervisor outlining the significance of its content.
- The theses should have a clearly defined structure including an introduction accessible to new PhD students and scientists not expert in the relevant field.

Indexed by zbMATH.

More information about this series at https://link.springer.com/bookseries/8790

Ali Aryo Bawono

Engineered Cementitious Composites for Electrified Roadway in Megacities

A Comprehensive Study on Functional Performance

Doctoral Thesis accepted by
Technical University of Munich, Germany

Springer

Author
Dr.-Ing. Ali Aryo Bawono
Department of Civil Geo
and Environmental Engineering
Technical University of Munich
Munich, Bayern, Germany

Supervisors
Prof. Dr.-Ing. Stephan Freudenstein
Department of Civil Geo
and Environmental Engineering
Technical University of Munich
Munich, Bayern, Germany

Prof. Dr. En-Hua Yang
School of Civil and Environmental
Engineering, Nanyang Technological
University Singapore
Singapore, Singapore

ISSN 2190-5053 ISSN 2190-5061 (electronic)
Springer Theses
ISBN 978-3-030-88541-0 ISBN 978-3-030-88542-7 (eBook)
https://doi.org/10.1007/978-3-030-88542-7

This Springer imprint is published by the registered company Springer Nature Switzerland AG
The registered company address is: Gewerbestrasse 11, 6330 Cham, Switzerland

Supervisor's Foreword

It is a great pleasure to introduce Dr.-Ing. Ali Aryo Bawono's doctoral thesis work, accepted for publication within Springer Theses and awarded with a prize for an outstanding original work. Dr.-Ing. Bawono joined my research group for Civil, Geo and Environmental Engineering at Technische Universität München in February 2017. He started his doctoral study with a 4-year scholarship from the National Research Foundation of Singapore through TUMCREATE research institute. He completed it with an oral defence on September 15, 2020.

The electromobility concept has now started to become a trend to achieve the ultimate public transport system in megacities. Along with electromobility, autonomous Electric Vehicles (EVs) manufacturers have started flooding the automotive market. Road infrastructure needs to be adapted and reinvented to support these positive trends. The concept of an electrified roadway with Dynamic Wireless Power Transfer technology can be a promising solution to charge Electric Vehicles (EVs) in megacities such as Singapore, Munich, London, New York, and Beijing. This concept offers many advantages, including the availability of continuous power and the possibility of using smaller battery capacities for EVs compared to those used conventional static charging.

Dr.-Ing. Bawono's thesis is a very comprehensive work reporting on the design of a pavement concept for an electrified roadway using a material called Engineered Cementitious Composites (ECC). Dr.-Ing. Bawono presents a modified Performance-Driven Design Approach (PDDA) to achieve the desired mechanical properties as well as to reach the objectives of the optimum design in terms of functional and structural performance for pavement made of ECC. The ECC is modified to fulfill the safety and comfort factors without neglecting the characteristics of conventional ECC. The modified high-performance ECC is exhibiting an extreme tensile strain capacity much higher than conventional concrete. It enables a slab pavement to work without steel reinforcement and the ability to embed a Dynamic Wireless Power Transfer technology to charge Electric Vehicles (EVs). Thus, the interference of the electromagnetic field actuated by the inductive charging can be avoided as this may lead to a major reduction of the charging rate efficiency. Additionally, it allows the reduction of the thickness of ECC slab pavement significantly. The functional performance for pavement made of modified ECC, including skid resistance, surface water drainage, and tire-pavement noise, has been investigated and designed. The modified PDDA approach is then strengthened with the numerical design approach. 3D Finite Element and Fluid Dynamics Models are employed to understand the mechanism of textures affecting the functional performance for pavement made of ECC. The validated models are able to predict the functional performance of ECC pavement. In addition, an environmentally friendly photocatalytic function for pavement made of ECC has also been investigated. Further, a Multi-Criteria Design Analysis (MCDA) is proposed to identify the optimum functional performance.

Dr.-Ing. Bawono's thesis offers a comprehensive account of significant novel scientific contributions, which have been individually published in top journals and presented at some internationally recognized conferences. The approach presented here enables the fabrication of a robust and safe precast system for an electrified roadway that can be installed within a few hours, with a great potential for future electromobility applications. Thus, besides addressing academic researchers, this book will also provide practical knowledge for the transportation industry and authorities.

Munich, Germany
June 2021

Dr.-Ing. Stephan Freudenstein

Parts of this thesis have been published in the following documents:

Journals

Bawono, A. A., Tan, Z. Hamdany, A. H., NguyenDinh, N., Qian, S., Lechner, B., Yang, E.-H. (2020). Bright and slip-proof engineered cementitious composites with visible light activated photo-catalysis property for pavement in tunnels. *Cement and Concrete Composites*, 114, 103387. doi: 10.1016/j.cemconcomp.2020.103788.

Bawono, A. A., Lechner, B., Yang, E.-H. (2019). Skid resistance and surface water drainage performance of engineered cementitious composites for pavement applications. *Cement and Concrete Composites*, 104, 103387. doi:10.1016/j.cemconcomp.2019.103387.

International Conferences and Workshops

Bawono, A. A., NguyenDinh, N., Lechner, B., Yang, E.-H. (2019). Study of finite element model on skid resistance for pavement made of Engineered Cementitious Composite. Structural Engineering, Mechanics and Computation, Cape Town, South Africa, September. CRC Press, London. DOI: 10.1201/9780429426506-275.

Bawono, A. A., Caliskan, E., Lechner, B., Yang, E.-H. (2019). Multi criteria analysis for the functional performance of surface pavement. Advances in Engineering Materials, Structures and Systems: Innovations, Mechanics and Applications, Cape Town, South Africa, August. CRC Press, London. DOI: 10.1201/9780429426506-284.

Bawono, A. A., Lechner, B., Alhaz, F. I. (2018). Asphalt pavement analysis with 3D FEM Model using actual tire pavement contact stresses measure with pressure mapping system device. Transportation Research Board 97th Annual Meeting, Washington DC, United States, January 2018.

NguyenDinh, N., Bawono, A. A., Yang, E.-H., Qiu, J. S., Lechner, B. (2018). Preparations for the field trial of preacast ultra-thin whitetopping (PUTW) in Singapore. in Proceedings of the 9th International DUT-Workshop on Research and Innovations for Design of Sustainable and Durable Concrete Pavements, Potsdam, Germany, Jun. 2018.

Bawono, A. A., NguyenDinh, N., Yang, E.-H., Lechner, B. (2017). Evaluation of highway pavement skid resistance performance made of Engineered Cementitious Composite (ECC). Tenth International Conference on the Bearing Capacity of Roads, Railways and Airfields, Athens, June. CRC Press, Greece. DOI: 10.1201/978131510 0333-134.

Patent and Technology Disclosures

NguyenDinh, N., Aguilar, O. A. C. A., Bawono, A. A., Lechner, B. (2020). Connection assembly, levelling or lifting system, and method of levelling slab. International Patent WO/2020/171775, 12 February 2020.

NguyenDinh, N., Bawono, A. A., Qiu, J. S., Yang, E.-H., Lechner, B, Ng, K. W. (2018). Polymer based mat (silicon) to create studded surface for precast concrete pavement slab to improve the interface bonding between the precast slab and its underneath layer. NTU TD No. 2018-263, September 2018.

Yang, E.-H., Qiu, J. S., NguyenDinh, N., Bawono, A. A., Lechner, B. (2017). Durable precast engineered cementitious composites ultra-thin whitetopping (PECCUTW) system for fast road paving. NTU TD No. TD/396/16, February 2017.

Acknowledgment

I would like to thank Prof. Stephan Freudenstein at the Technical University of Munich (TUM) in Germany for his expert advice and guidance throughout this challenging research. His extensive experience and knowledge in infrastructure engineering were really helpful in getting me the right direction for my research. Prof. Freudenstein has been very supportive since day one to the final stage of my research.

I would like to express gratitude to Prof. En-Hua Yang at the Nanyang Technological University (NTU) in Singapore for his continuous guidance. Prof. Yang has been a constant and great source of knowledge and playing a big role in my research from the material science perspective. I am grateful to learn how to balance between doing good research and presenting research in the journal to benefit communities.

I would like to acknowledge, with special gratitude, Dr. Bernhard Lechner for all his advice, energy, enthusiasm, and encouragement. I might not have started the doctoral study and finally be in this stage without his important roles. Through extensive mentoring, I have learned to understand the essence of the research and embrace it.

I am thankful to PPI Prof. Fritz Busch, PPI Prof. Wong Yiik Diew, and PI Dr. Andreas Rau, who are always supporting and giving me the responsibility and freedom to carry out the research, and to create a very positive environment in TUMCREATE.

I am also thankful to all of my friends, colleagues, and students in the RRT Department of TUMCREATE (Nen, Teron, Zain, Xiaodong, Freddy, Mike, Meng, Emmeline, Christian v.S), the BGU TUM (David, Sophie, Matthias, Kangle, Martina, Norbert, Yang, Christian, Dr. Simon, Dr. Geisler, Alhaz, Emre, Janaki), and the Living Stone Research Group (Jishen, Zhitao, HeShan, Krishnan, Junxia, Murugan, Cem, Jiawei, XiaoXi, Johnson). Countless guidance, help, support, and motivation I got from all of you always made me feel blessed to be a member of this big family.

The research project done in collaboration between two universities TUM and NTU was an amazing experience for me. I had never really felt unnecessary pressure, but instead, I found two amazing, humble, and warm families who might have different principles. Still, both have similar spirits for the greater purpose in science and education.

I am grateful to my beloved Mama and Papa (may you have peace in heaven) for raising and educating me. I would like to say thank you for the endless support, love, and understanding from my lovely Desi Adiyati and Alia Alessandra Bawono.

My humble gratitude is due to Ir. Subagyo, CES and Dr. Ir. Hedy Rahadian, M.Sc for their continuous supports.

This work was financially supported by the Singapore National Research Foundation under its Campus for Research Excellence and Technological Enterprise (CREATE) programme.

Munich, Germany
June 2021

Dr.-Ing. Ali Aryo Bawono

Contents

Symbols and Abbreviations

Symbols

Vf	volume fraction of fiber
lf	length of fiber
df	diameter of fiber
Ef	elastic modulus of fiber
σfu	tensile strength of fiber
Km	fracture toughness of ECC matrix
Em	elastic modulus of ECC matrix
a_0	initial flaw size distribution of ECC matrix
τo	interfacial frictional stress of interface
GD	chemical bond of interface
f	snubbing coefficient of interface
f'	strength reduction factor of interface
β	slip-hardening behavior of interface
SN	skid number
F	tractive force, N
W	dynamic vertical load on a test wheel, N
MuN	Mu Number
BPN	British Pendulum Number
Sp	speed number
a,b	calibration constants
TX	macrotexture (MPD or MTD) measurement, mm
$F(60)$	friction number
$FR(60)$	adjusted value of friction measurement at selected slip speed S to a slip speed of 60 kilometers per hour
λ	wavelength
A	amplitude
μ	friction level
MTD	mean texture depth of pavement macrotexture, mm

V	sample volume, mm^3
D	average diameter of the area covered by the material, mm
$S\text{x}$	cross slope, m/m
S	longitudinal gradient, m/m
$S\text{r}$	slope of resultant flow path, m/m
$L\text{f}$	length of flow path, m
L	pavement width, m
WD	water depth above the top of the surface asperities, mm
WFT	water film thickness, mm
TF	length of flow path, m
TXD	texture depth, mm
I	average rainfall intensity, mm/hour
$\text{S}r$	slope of resultant flow path, m/m
Qr	peak runoff at the point of design, m^3/second
C	runoff coefficient
I	average rainfall intensity, mm/hour
A	catchment area, hectares
w	relative weight of importance
x	performance value
n	decision criteria

Abbreviations

AAC	Autoclaved aerated concrete
APTM	Accelerated Pavement Tracking Machine
BPN	British Pendulum Number
BPT	British Pendulum Tester
COP	Code of Practice
CPB	Controlled pass-by (noise test)
CPX	Close Proximity (noise test)
CTIM	Continuous flow traffic time-integrated model (noise test)
DART	Dynamic Autonomous Rapid Transit
DFT	Dynamic Friction Tester
DOT	Department of Transportation
DPT	Dynamic Power Transfer
DPUR	Dirt Pick-Up Resistance
DWPT	Dynamic Wireless Power Transfer
ECC	Engineered Cementitious Composite
ECC-Cor	Engineered Cementitious Composite mixed with corundum
ECC-SS	Engineered Cementitious Composite mixed with silica sand
EPA	Environmental Protection Agency (of United States)
FEM	Finite Element Method
FHWA	Federal Highway Administration, agency of USA

HPFRCC	High-Performance Fiber Reinforced Cementitious Composites
IPT	Inductive Power Transfer
IRI	International Roughness Index
JRCP	Jointed Reinforced Concrete Pavement
JTC	Jurong Town Corporation (Singapore State-Owned Company)
KAIST	Korea Advanced Institute of Science and Technology
LCCA	Life Cycle Cost Analysis
LTA	Land Transport Authority (of Singapore)
LWAC	Lightweight aggregate concrete
MPD	Mean Profile Depth
MTD	Mean Texture Depth
NEA	National Environment Agency (of Singapore)
NTU	Nanyang Technological University (of Singapore)
NWC	Normal Weight Concrete
OBSI	On-board sound intensity
OFT	Outflow Meter Test
OLEV	On-Line Electric Vehicle
PAC	Portland Asphalt Concrete
PCC	Portland Concrete Cement
PDDA	Performance-Driven Design Approach
PFA	Pulverized fuel ash
PrI	Profile Index
PSR	Present Serviceability Index
PUB	The Public Utilities Board (of Singapore)
PUTW	Precast Ultra-Thin Whitetopping
PV	Photovoltaics
PVC	Polymerizing Vinyl Chloride
PZT	Piezoelectric transducer
RhB	Rhodamine B
ROW	Right of Way
SCRIM	Sideway-force Coefficient Routine Investigation Machine
SDRE	Standard Details of Road Elements
SHCC	Strain Hardening Cementitious Composites
SP	Superplasticizer
SPB	Statistical Pass-by (noise test)
SPERoN	Statistical Physical Explanation of Rolling Noise
SPM	Sand Patch Method (test)
SRT	Semi-rapid Transit
TRL	Transport Research Laboratory
TUM	Technical University Munich (of Germany)
TUMCREATE	Technical University Munich and Nanyang Technological University join forces as part of the Campus for Research Excellence And Technological Enterprise (CREATE)
TWT	Thin Whitetopping
UHI	Urban Heat Island

URA	Urban Redevelopment Authority (of Singapore)
UTW	Ultra-Thin Whitetopping
UV	Ultraviolet (Light)
VL	Visible Light
VMA	Viscosity Modifying Agent
XPS	X-ray photoelectron spectroscopy

Chapter 1
Introduction

In this chapter, the motivational background, the objective, and the outline of the research is explained in detail. Nowadays, safety is becoming as one of the essential factors in road design, which cannot be neglected. Road designers (and roadway authorities) are raising the bar in designing the roadways. The roadways, including electrified roadways, should be operationally efficient and comfort, structurally stable, and importantly provide safety for the driving experience.

1.1 Research Background

In recent years, the concept of an electrified roadway with wireless inductive charging for electric vehicles (EVs) was proposed as a solution for electromobility to achieve an ultimate public transport system. In this thesis, the research is focusing on the electrified roadway with a precast method that can be implemented in megacities such as Singapore, to reduce the construction time and to minimize the road closure duration. A new pavement system named Precast Ultra-thin Whitetopping (PUTW) is developed for rapid retrofitting and conversion of existing road infrastructure into electrified roadways and or conventional roadways. This thesis is tackling not only the structural performance aspect but also the functional performance aspect which commonly ignored.

Engineered Cementitious Composite (ECC) is a high-performance fiber-reinforced cementitious composite exhibiting extreme tensile strain capacity of several hundred and flexural strength 2–3 times that of normal concrete. The usage of ECC material enables the design of PUTW for an electrified roadway with the thickness of the slab is greatly reduced. And the slab can be produced without the need for steel reinforcement as reinforcement can cause interference of electromagnetic field of inductive charging and lead to the reduction of charging efficiency. However, due to the omission of coarse aggregates in ECC mix design, pavement made of ECC generally shows low skid resistance, which might cause skid-accidents in highway

A. A. Bawono, *Engineered Cementitious Composites for Electrified Roadway in Megacities*, Springer Theses, https://doi.org/10.1007/978-3-030-88542-7_1

traffic. On top of that, Singapore experiences a high intensity of rainfall. Almost half of the year is rainy days with hourly average precipitation of 120 mm, increasing the risk of hydroplaning, leading to a wet-skid accident.

The functional performance of the pavement is based on its ability to serve the user, which is based on significant indicators such as roughness, skid resistance, risk of hydroplaning, appearance (texture), and other safety considerations. These indicators are related to the surface properties of the pavement, which is identified as texture [1]. In this thesis, laboratory and field tests were conducted to get a better perspective of safe pavement. Simulation using numerical models is also performed. Simulation with finite element method is used to broaden the understanding by extending the scale of pavement system used in the laboratory. Other valuable data are obtained by field tests, which also used to validate the numerical calculation.

In recent years, roadway authorities have considered more functional parameters when designing the surface properties of the pavement. The geometrical aspect, durability, workability, noise, brightness and self-cleaning properties for electrified roadway made of ECC, which have never been studied before, are addressed in this thesis.

A perfect solution that maximizes all these parameters might be challenging to achieve. Yet, an optimum solution to make a safe electrified pavement can be identified with the help of the Multi-Criteria Decision Analysis (MCDA) tool. In this thesis, MCDA is developed to identify the optimum solutions. Those criteria then are being scored and weighted. Finally, the results are visually presented with the spider chart.

1.2 Aims and Scope of the Research

The main motivation behind this thesis is to improve the performance of electrified roadways through enhancing the safety and imparting additional functions into the electrified roadway (PUTW with ECC material). Ultimately, the electrified roadway can provide a robust system that not only reliable on the structural but also functional performance. The scope of this research included the following objectives:

- To study the state of the art of electrified roadways, from the functional performance (surface pavement properties) perspective.
- To identify the optimum solution of the functional performance for surface pavement by developing Multi-Criteria Decision Analysis.
- To develop material for an electrified roadway which fulfils the functional performance of safety requirement. The requirements include the studies of skid resistance and surface water drainage performance. The scope of the study covers laboratory test, field investigation, and simulation using numerical tools.
- To further develop green material for an electrified roadway, which has the function of low noise, high solar reflectance to reduce urban heat island, as well as self-cleaning capabilities to maintain the high reflectance.

1.3 Research Outline

This thesis focuses on the design of the functional performance of electrified roadways by using an innovative Engineered Cementitious Composites (ECC) material and developed with the precast method. The content of the following chapters is briefly introduced here:

- Chapter 2 examines the current state of the art in electromobility of Singapore. It starts with a general introduction regarding the transportation vision done by TUMCREATE for Singapore. It also describes the current road traffic, the road emission condition, and the conventional pavement design in Singapore. Lastly, it highlights the state of the art of electrified roadways.
- Chapter 3 examines the current state of the art in the structural design of ultra-thin whitetopping. The chapter highlights the concept design of ECC material. Further, the chapter describes the slab manufacture and construction processes.
- Chapter 4 gives an overview of the functional performance of pavements. It focuses on a few parameters, including skid resistance, surface water drainage, tire-pavement noise, durability, and photocatalytic functions.
- Chapter 5 presents the collection of functional aspects including microtexture, macrotexture, surface water drainage, noise, pavement roughness, cost, and durability. This chapter also presents the proposed methodology and analysis of the Multi-Criteria Design Analysis (MCDA) on selecting the optimum design for roadways based on many criteria. The MCDA can be adopted and then used by roadway designers and authorities to select the optimum design for their roadways and or electrified roadways.
- Chapter 6 explains the study on the mechanical properties of the ECC material. It includes the methodology of the experimental tests on how the ECC can be modified with the modified Performance-Driven Design Approach (PDDA) method. Thus, the ECC can fulfil both structural and functional performance aspects required for electrified roadways.
- Chapter 7 presents the skid resistance laboratory tests conducted with British Pendulum Tester (BPT) and Sand Patch Method (SPM) tester to measure the microtexture and macrotexture, respectively. The tests were done on the improved ECC mixture design with few combinations of material, including corundum and silica sand. It also describes various methodologies on the macrotexture improvement by introducing various grooves.
- Chapter 8 presents the skid resistance tests on the full-size PUTW slab applied on several applications, including bus-bay and pedestrians. The PUTW slab is made of ECC mixed with corundum material with different finishing methods, including exposing and brushing.
- Chapter 9 studies the skid resistance on the pavement made of ECC material with a numerical method. The numerical studies were done based on a three-dimensional (3D) finite element method (FEM) model by using a rigid dynamics model with ANSYS software. The 3D FEM model has shown a good validation with the

laboratory tests. This model aims to predict the skid resistance performance for pavement made of ECC with various textures.

- Chapter 10 presents the water surface drainage laboratory tests conducted with the water flow meter to measure how fast the water removed from the ECC surface pavement. It also described the test with tailored design apparatus on measuring the water film thickness on the ECC surface pavement. The tests were conducted on the improved ECC mixture design mixed with corundum material with various grooves. This chapter also presents the study on the effect of pavement geometry on water surface drainage.
- Chapter 11 presents the water surface drainage field test on full-size PUTW slab applied on several applications including bus-bay and pedestrians. The PUTW slab is made of ECC mixed with corundum material with different finishing methods, including exposing and brushing.
- Chapter 12 describes the numerical study that employed by using ANSYS Fluent (Fluid Dynamics Analysis). The numerical study consists of the development of the 3D model for the simulation of surface water drainage. The 3D model includes the geometry development, mesh design, material properties, conditions, model validation and analysis. The 3D FEM model has shown a good validation with the laboratory tests. The objective of the model is to simulate how surface texture and geometry of pavement made of ECC will affect water surface runoff.
- Chapter 13 explains the proposed correlation mathematical model between the skid resistance and the surface water drainage performance based on analysis from data collected and numerical simulation modelled. Further, the mathematical model also takes the pavement geometry aspect into it. This fact emphasizes that those two criteria should be designed well that they are contributing to the safety comprehensively.
- Chapter 14 presents the innovative contact-free measuring test applied to the specimens, allows for a non-destructive determination of the macrotexture of the slab surfaces. The results are then simulated by using SPERoN software to predict the tire-pavement noise. The modification of macrotexture design on the ECC slab has been successfully reduced the tire-pavement noise that met with the limit set by roadway authorities.
- Chapter 15 reports the development of a bright and slip-proof ECC with visible light activated photocatalysis property for pavement in tunnels. Different T_iO_2 nano-particles were tested first to evaluate their photocatalysis property under visible light irradiation. The selected T_iO_2 nano-particle was used to develop the bright and slip-proof ECC with the inclusion of white fine aggregates. Effects of fine aggregate content on the mechanical properties and skid resistance of the resulting ECC were investigated. Influences of T_iO_2 dosage on the mechanical properties of ECC were tested and self-cleaning performance and dirt pick-up resistance (DPUR) of ECC under visible light irradiation were studied.
- Chapter 16 summarizes the accomplishments of this thesis, including key contributions and findings. It also describes the proposed future research directions.

Reference

1. AASHTO (1993) Design of pavement structures. AASHTO, Washington, D.C.

Chapter 2
Electro-Mobility Solution Towards the Ultimate Public Transport System with a Case Study of Singapore

This chapter has reviewed the state of the art of road transport in Singapore and how the current gap leads to the needs of an innovative solution to achieve the ultimate public transport system in Singapore. An electric taxi concept called EVA for Singapore environment was developed. A new system, Dynamic Autonomous Rapid Transit (DART) which will close the transport gap between Mass Rapid Transport (MRT) and Bus is also introduced. Both EVA and DART are developed based on electrification concept. This leads to an alternative solution for future pavement design, the electrified roadway concept.

2.1 Towards the Ultimate Public Transport System for Singapore

Singapore is often used as the benchmark in electromobility concept by other megacities. The concept of electromobility in megacities is introduced with Singapore as a case study [24]. Singapore's electromobility concept used the traffic and power infrastructure in the assessment, which was then used to develop and introduce EVA, its electric taxi. Based on the driving pattern in Singapore, EVA can travel up to 200 km with only 15 min of charging. This electric taxi fleet is to replace the fuel-burning taxis that will help to reduce emissions and noise. However, it was eventually concluded that manufacturing EVA in Singapore is not feasible with one of the reasons is cost.

The electromobility concept was leveraged to a higher level with a new objective that aims for better comfort and efficiency for commuters to achieve the ultimate public transportation system in Singapore. The current system has a transport gap between the MRT and bus network that contributes to that less efficient system and therefore sacrifice the comfort of the commuters. The new proposed system, called Dynamic Autonomous Rapid Transit (DART), will bridge the two systems. DART, as stated in its name, will utilize electric vehicles with autonomous capability.

A. A. Bawono, *Engineered Cementitious Composites for Electrified Roadway in Megacities*, Springer Theses, https://doi.org/10.1007/978-3-030-88542-7_2

Fig. 2.1 **a** The electric taxi EVA **b** Dynamic Autonomous Rapid Transit system

DART will use a modular powertrain for the energy storage solution. DART is a 30-seater unit with dimensions of 6 m $\times$ 2.7 m $\times$ 2.9 m with a laden weight of approximately 7000 kg. As can be seen from Fig. 2.1, the operation can be running on platooning mode with up to 10 modules connected by electronic coupling. The maximum speed is 78 km/h with an electronic limit of 60 km/h per Singapore traffic regulation. When it is fully charged, the DART with a 120 kWh battery will be able to provide public transport service for an estimated more than 8 h, covering 192 km. DART operation will be autonomous, mainly to reduce the accident rate. In the usual case, DART has a higher level than private vehicles, so that private vehicles need to give way to DART. The prioritization will be enabled by virtual right of way (V-ROW).

Static charging stations implementation was then analyzed to maintain the state of charge of the usage of UVs in Singapore. Placement of charging stations with fast charging performance was proposed only at locations where EVs are likely to run out. Alternatively, charging stations can be placed where ECs started or ended their trips. At the same time, TUMCREATE was also conducting research on innovative pavement system based on local conditions. The proposed system provides possibilities for the implementation of electrified roadways as well as enables fast replacement for maintenance with a long life-cycle.

2.2 Road Traffic in Singapore and Its Emissions

Singapore has a total population of 5.5 million in 2016, within an area of 700 Km^2. Due to land scarcity, Singapore embraces public transport to achieving long term plan transport sustainability. Many challenges are faced, including increasing travel demand, limited land, declining public transport modal share, changing demographics with their expectations for better transportation. Three strategies are planned to tackle those challenges: firstly, by enhancing the quality of public transport, managing road usage, and meeting diverse needs [14]. The daily ridership of public transportation based on roadways in comparison to railway based is typically higher. LTA had added more than double the number of premium or high-quality bus services

Table 2.1 Motor vehicle population in Singapore

Motor vehicle population	2010	2014	2015
Cars	584,399	600,176	575,353
Rental cars	13,347	18,847	29,369
Taxi	26,073	28,736	28,259
Buses	16,309	17,554	18,183
Motorcycles	148,160	145,026	143,900
Goods & others	157,541	161,698	162,182
Total	945,829	972,037	957,246

annually. For example, bus services have been increased to 91 bus services per January 2013 compared to only about 42 bus services in 2008 [14]. Reciprocally, in 2014, the total number of average daily ridership for MRT and LRT has doubled in the last ten years, with 2900 passenger trips only. While, during the same period, bus and taxi cover 4700 passenger trips [7]. Singapore is maintaining the number of private vehicles in a specific range of numbers to keep the balance of the modal shares. There is no significant increase in the motor vehicle population from 2010 to 2015. From 2010 to 2014, the vehicle population increased 2.8%, and then the population declined 1.2% in the following year, caused by the declination number of cars and motorcycles with 4.1% and 0.8%, respectively. Only buses as public transport vehicles are increased from 2014 to 2015, with 3.6% incremental. Further details about the motor vehicle population are summarized in Table 2.1 [15].

A statistic issued by Singapore Land Transport Authority (LTA) [15] states that in 2021, Singapore has a total length of the paved road approximately 3,425 km (9080 lane-kilometers) approximately which covers 161 km (1059 lane-kilometers) of expressways, 652 km (3054 lane-kilometers) of arterial roads, 561 km (1584 lane-kilometers) of collector roads, and 2051 km (3381 lane-kilometers) local access roads. In 2015, the number increased to 9264 lane-kilometers [5], as shown in Fig. 2.2. This is resulting in a road density of 5 km/km^2 approximately.

Singapore contributes around 0.11% of global emissions. In 2015, Singapore ranked 123rd of 142 countries in CO_2 emissions per dollar GDP [17]. It is projected that the Business-as-usual (BAU) emissions in Singapore are expected to reach 77.2 million tons in 2020 [17]. Balakrishnan cites the Sustainable Singapore Blueprint 2015 that Singapore is aiming to reduce its carbon emissions by up to 11% below business-as-usual-scenario levels by 2020 [20].

2.3 Current Pavement Design in Singapore

Current Singapore pavement design is standardized in the Standard Details of Road Elements (SDRE) issued by LTA. Three types of pavement designs are currently

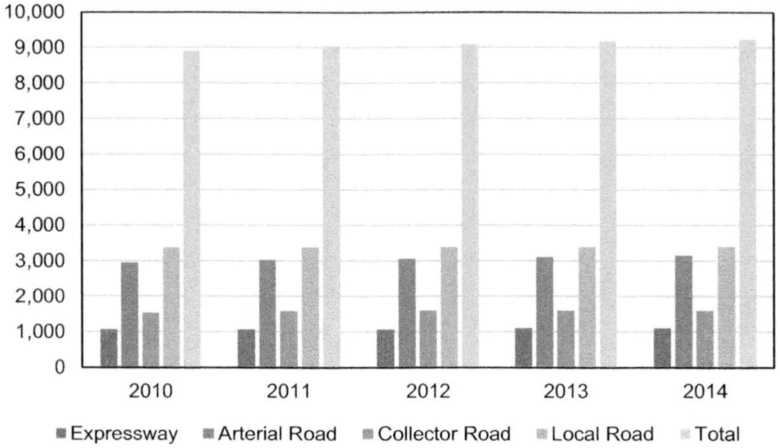

Fig. 2.2 Length of roadway (Lane-Kilometers) in Singapore

applied in Singapore: asphalt pavement design, concrete pavement design, and semi-rigid pavement design. Asphalt pavement is typically used on the carriageway, while concrete pavement and semi-rigid pavement are typically applied at bus bays and intersections.

2.3.1 *Asphalt (Flexible) Pavement*

Typically, asphalt pavement in Singapore consists of four layers: sub-base, base course, base course, and asphaltic wearing course. However, an additional layer with porous asphalt is also introduced to reduce the noise. LTA introduces four different types of asphalt pavement: Type 1A, 1B, 2, and 3 (Land Transport Authority 2015). Each type has different thickness and material properties, and its usage depends on the road type. The illustration is shown in Fig. 2.3, and explanations are elaborated in Table 2.2.

Fig. 2.3 Asphalt (flexible) pavement structure design

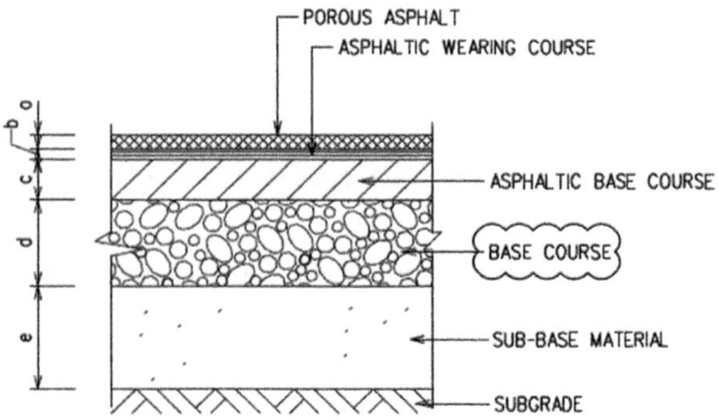

Table 2.2 Type of asphalt (flexible) pavement

Type of flexible pavement	Road type	Thickness				
		a	b	c	d	e
Type 1A (with porous asphalt)	Expressway	40	30	120	250	300
Type 1B (without porous asphalt)	Semi-Expressway	0	50	120	250	300
	Arterial road					
	Road in industrial area					
Type 2	Primary access	0	40	90	200	300
Type 3	Local access	0	25	75	200	200

2.3.2 Concrete (Rigid) Pavement Design

LTA publishes rigid pavement design not only for road carriageway but also for bus bay and intersection applications. Rigid pavement used in Singapore is typically a jointed reinforced concrete pavement (JRCP) type with a thickness of 225 mm. Typically, using the base course as an unbound support layer supported with a geotextile membrane to separate between each layer. Rigid pavements can be used altogether with asphalt pavements for carriageway. In addition, three meters transition slab is needed to provide a gradual transition. A bump might appear over a period of time on the AC pavement side if a transition slab is not used [12]. The picture is shown in Fig. 2.4.

In Singapore, rigid pavement is also used on bus bays and intersections to handle the higher stresses due to braking force and acceleration force, particularly from heavier vehicles such as buses and trucks. A slight transition is longitudinally designed on the bus bay, where 105 mm thick of the concrete of the bottom part is extruded into the asphalt pavement with a width of 150 mm. See Fig. 2.5a, b for details.

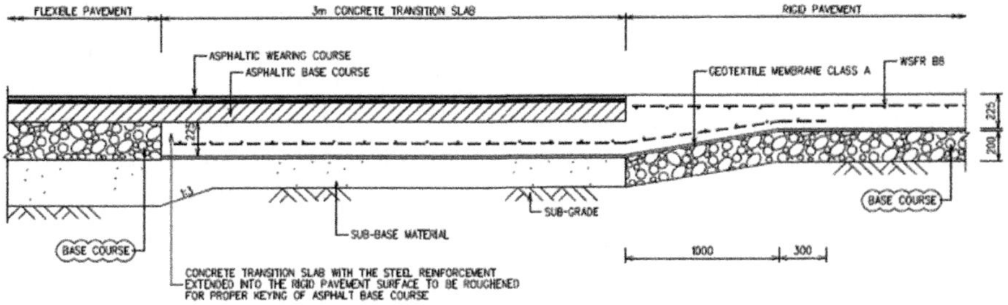

Fig. 2.4 Rigid pavement and its connection with flexible pavement

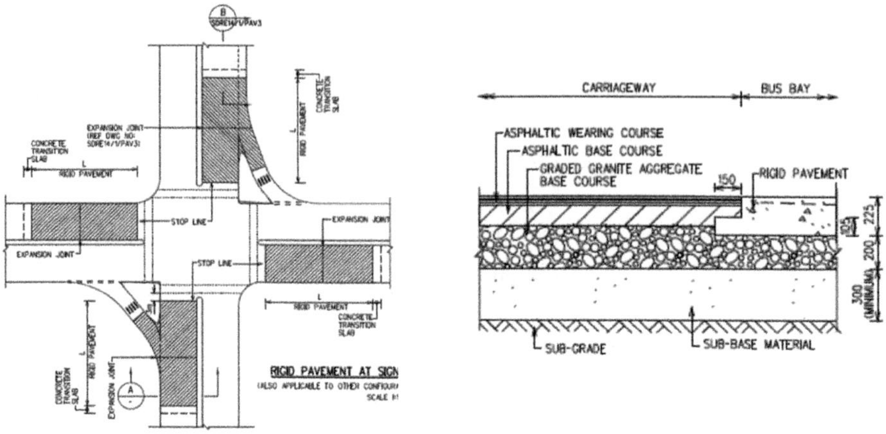

(a) (b)

Fig. 2.5 **A** top view of pavement design at intersection, **b** structure for bus bay

2.3.3 Semi-Rigid Pavement Design

In Singapore, significant application of the semi-rigid pavement is for traffic light junction with a typical thickness of 50 mm as the wearing layer [25]. Semi-rigid pavement is designed as an open-graded asphalt with a void between 25–30% by volume (Marshall) filled with high strength cementitious mortar [13]. It is basically the combination of Porous Asphalt Concrete (PAC) and Portland cement concrete (PCC) pavements. The semi-rigid pavement has to achieve a compressive strength of 7–10 MN/m^2 at 28 days and flexural strength of approximately 3.5 MN/m^2. High strength cementitious mortar consists of a dense packing of cementitious inorganic material mixed with water. The mortar needs to achieve a compressive strength of 110 MN/m^2 at 28 days and flexural strength of approximately 15 MN/m^2. The mix design should use a water to binder ratio within a range between 0.226–0.278 [13]. Some advantages of semi-rigid pavements include higher resistance to rutting than asphalt concrete, relatively easier to construct and maintain, and faster in respect of open to traffic, which can be done less than 24 h. Therefore, filling high strength cementitious mortar should be done carefully to avoid excessive material on the surface. The excessive material can cover the aggregates, which may cause lower skid resistance. It is widely acknowledged that semi-rigid has a longer service life span than asphalt concrete but less than PCC. However, semi-rigid has its disadvantages include surface-initiated cracking (top-down cracking and thermal fatigue cracking), instability rutting, fatigue cracking of cement-treated materials, shrinkage cracking and reflective cracking [8].

2.4 Future Pavement Design: Electrified Roadways

Vehicle technologies are significantly advancing that help to fight against the emission. As technology develops, ultra-low emission vehicles (ULEVs), including pure electric vehicles, plug-in hybrid and fuel cell electric vehicles, are trending nowadays. Those vehicles are available on the market and will be a common sight in Singapore in the coming years. At the same time, transport infrastructures are also developing to not only helping to reduce the emissions but also to allow safer journeys for the vehicles.

Currently, two technologies are available to charge electric vehicles' battery: plug-in charging and inductive charging. The plug-in charging system operates at a specific supply point. It requires a cable and a socket that need to be connected to the vehicle. Inductive charging works differently. This system, commonly known as electrified roadways, provides a charge to electric vehicle batteries without any physical connection, allows charging the vehicle when the vehicle is stationed (static charging) or on the move (known as dynamic charging). Dynamic charging, also called opportunity charging, results in an increased operating time by two to three times, allows reducing the battery costs and weighting per vehicle as well as space for the battery. It reduces the overall reducing life cycle cost [11].

2.4.1 Development of Electrified Roadway

The installations of a wall power outlet for the charging system have been commonly completed in urban environments. However, there is a requirement for the dynamic charging system, which embedded in the pavement, that will enable electric vehicles to travel greater distances on the strategic road network by using only electric power [9]. This system is often called electrified roadway. A recent article explains that the adoptions of electric vehicles are still remaining low due to the fact that the battery has many issues such as high price, heavy, and it offers a low range of distance [22]. Surveys show that EV users or EV buyers have concerns with the infrastructure of the EC charging, specifically the distribution and the time needed to charge their EVs. A similar hypothesis is given by the Guardian [16] with the conclusion there is a need to have an extension of charging infrastructure.

The electrified roadway concept was firstly invented for railway application by Hutin and Leblanc [10] in 1894. In 1970, a 6 kW proof of concept was tested at Lawrence Berkeley and Lawrence Livermore National Libraries. However, the conclusion drawn from tests shows that the system was operationally not suitable. A similar outcome was given from the Santa Barbara Electric Bus Project (1979) consists of a planning and feasibility study followed by preliminary engineering and prototype development and testing.

A project called Roadway Powered Electric Vehicle (RPEV) was done by California Partners for Advanced Transit and Highways (PATH), Institute of Transportation Studies, UC Berkeley, and the Southern California Association of Governments (SCAG) in the 1980s. It is then recognized as the first successful test in a dynamic-type charged pavement [3, 21]. Within the project, approximately 200 m road track was constructed in their facility. The track consisted of two 60 m electrified segment. An electric bus of 35 passengers with a maximum speed of 64 km/h was used. Details of RPEV components can be seen in Appendix A.

Korea Advanced Institute of Science and Technology (KAIST) developed a shaped magnetic field in resonance (SMFIR) in 2009 to enable wireless electric power transmission for moving vehicles. A total length of 372.5 m of powered track of electrified roadway was built in Seoul Grand Park. The system is operating in 2013 on a 25 km line in Gumi. It is reported that the advantage of KAIST system is that the battery of the rechargeable bus is only 1/5 the size of a normal electric bus battery.

In 2009, Bombardier introduced PRIMOVE offering a wireless power supply for all vehicles. A testbed of 600 m was built at Augsburg. It is using a power supply of 750 DC. In 2010, Bombardier continued their research by demonstrating another testbed in Lommel for a bus. The testbed has 125 m in length built on concrete roads. Then recently, in 2015, the PRIMOVE Mannheim project has been developed for bus line 63 along nine kilometers. The system consists of four charging stations with an approximate charging time of two minutes.

In 2015, an executive agency of Highways England published a feasibility study with the topic of powering electric vehicles on England's major roads [23]. It is reported that the electrified roadway testbed consists of a track length of approximately one km with at least two lanes, each of 3.5 m wide each lane. Based on their investigation on road construction methods, the in-situ full-width lane was the preferred option for the off-road trial and suitable on highways.

The first charging rail-based electric roadway installed in Sweden in 2017 through the eRoadArlanda project with two kilometers constructed between Arlanda and the Rosersberg logistic area. The electrified roadway was mainly intended to be used by logistic trucks. However, this is not the first time Sweden has a pilot project for such a system. Back in 2012, they had a testbed with a total length of 200 m. An additional 150 m of electric rails were then installed in 2014. It is then followed by another 50 m of rail installed with the new generation of rail tracks [6].

2.4.2 Dynamic Power Transfer (DPT)

Research institutes and industries are currently developing new technology that allows power transfer to the electrified vehicle powertrain, including hybrid powertrains. A vehicle can use a power transfer system to charge the on-board Rechargeable Energy Storage System (RESS) or provide power to the electric motor.

Typically, such power transfer systems are plug-in electric chargers that charge the vehicle batteries at varying levels of power while the vehicle is stationary and switched off. However, it is also possible to use Wireless Power Transfer (WPT) to charge the batteries while stationary, using charging "pads". Both solutions are adequate for charging at home or in public parking areas but still require the vehicle to stop in an appropriate location to charge the battery. Inductive Power System (IPT) or Dynamic Power Transfer (DPT) is another option for supplying power to electric vehicles. It can be used while the vehicle is moving. It can help reduce or eliminate issues with a restricted range. DPT can be either conductive or wireless. According to the Transport Research Laboratory, the required power transfer level from a Dynamic Wireless Power Transfer (DWPT) system on a motorway is around 20–40 kW per vehicle for cars and light vans, and between 100–180 kW for trucks and coaches, based on the power required to maintain constant motorway speed [2, 23]

A few different institutions set standard and regulations for inductive charging for EVs. One of the first standard is ICNIRP 1998/2010L Guidelines for limiting exposure to time-varying EM fields (ICNIRP 1998), which regulates requirement related to Living Tissue affected by Power Dissipation caused by Electromagnetic Fields, a Limitation of Human Body SAR (Specific Absorption Rate, [W/kg]) by Limiting Electric and Magnetic Fields, and the distinction between "General Public" and "Occupational Exposure". Another standard was set in 2013 by SAE J2954 Wireless Charging Standard that explains the baseline for the safety consideration. It is stated that the common operating frequency is 85 kHz, the minimum charging efficiency should be at least 90%, parameters regarding interoperability, safety features, and validation methods, and as well as the charging levels are divided into three levels (1) WPT1 private low power is up to 3.7 kW, (2) WPT2 private and public space is up to 7.7 kW, and (3) WPT 3 for fast charging is up to 22 kW.

One of the first inductive charging systems was introduced in RPEV technology. It consists of the Inductive Coupling System (ICS), which transfers energy from a coil buried beneath the pavement surface to the vehicle through a magnetic field. In the RPEV system, the power conditioner used is an inverter with a three-phase 60 Hz power and then puts out a single phase at a higher frequency. The system allowed peak output by up to 60 kW to be transferred from the electrified roadway to the moving electric bus with an efficiency of up to 60%. Eventually, the system improved the electric bus range to 400 km, which is significantly higher than the previous performance of only up to 30 km.

Bombardier is one of the leading companies in WPT technology. They can provide an inductive charging system of up to 200 kW. They have applications that are widely used in Europe. While among research groups, KAIST shows a high interest in continuously developing the system with significant progress. Up to now, they successfully improved and implemented the system in Korea. According to the results reported by KAIST, the new wireless electric power transfer technology named Shaped Magnetic Field in Resonance (SMFIR) shown good performance with the efficiency of up to 85% with an air gap of 20 cm between the pavement surface and the pick-up installed on the vehicle [1]. The Wireless Power Transfer (WPT) systems are summarized in Table 2.3, and it is listed from the highest to the lowest power level.

KAIST has been developing OLEV for over a decade now. The first successful demonstration of the SMFIR system was done in 2009 that is identified as the 1st Generation (1G). Until the 4th Generation, the rail width is relatively wide. Then, they are successfully reduced from 140 cm (1G) to 80 cm (2G, 3G, and 4G). Only until the I-type is introduced, they can have a rail width with of only 10 cm. This narrow power supply with a width of 10 cm provides an overall power efficiency of 74% (including an inverter) at an output power of 27 kW and a large air gap of 20 cm. The electromagnetic field (EMF) is reduced significantly, with only less than ten mG. While S-type only needs 4 cm rail width yet producing 71% power efficiency at an output power of 22 kW. Details of OLEV development can be seen in Appendix A.

Table 2.3 Summary of WPT technology from different group

System	Air gap	Power level	Efficiency	Application and demos
EATON HyperCharger	30.5 cm (12 inch)	200 kW to 1 MW	–	Electric buses US targets: Tallahassee FL, Stockton CA, and Worcester
Bombardier PRIMOVE	4 cm	200 kW high power	95%	Electric buses and LRVs Germany and Belgium, 2011–2013
OLEV SMFIR KAIST	20 cm	100 kW	Up to 85%	Electric bus and LRVs KAIST and Gumi, Korea, 2009–13
PATH	6 cm	60 kW	60%	Electric bus
Technology (former Conductix-Wampfler)	4 cm	60 kW module in roadway,30 kW modular onboard bus	>90%	Electric Buses Turin and Genoa, Italy; Utrecht, Netherlands Utrecht since 2010; Lucerne, Switzerland; Lörrach, Germany; Rotorua, New Zealand
WAVE	20 cm	50 kW for 40 ft. bus & up to 140 kHz;	90%	Electric Buses USU, Logan, Utah; Monterey-Salinas Transit (MST); 2012–2014
Stanford	–	10 kW	>97%	Electric vehicles
ORNL	–	7 kW	90–94%	Electric vehicles
Siemens	15 cm	3.6 kW	>90%	Electric vehicles
WiTricity	15–200 cm	3.3 kW	>90%	Small electric vehicles

OLEV has developed an innovative power supply with its latest product. The new proposed S-type's rail width is smaller, but it also has another advantage which the power supply rail adopts a module concept. This concept makes it easier to fold itself by virtue of flexible thin power cables. Therefore, no power cable connection is required after being deployed [4]. With only 4 cm width is needed for the S-type power modules, this leads to easier construction work, relatively lower construction cost and faster deployment time. The modules can be embedded in asphalt or concrete pavement.

The technology of the inductive charging system is designed not as a replacement for the static system. Instead, it is to enhance the static system by having it in an integrated system. Both systems, static and dynamic charging, have their strength and weakness. The static system is relatively cheaper than the dynamic. But it requires the vehicles to be parked. Therefore, the system is only feasible to be placed on fixed station wise such as at home, at the office, at the shopping mall, or at a charging station. On the other hand, the dynamic system is more expensive than the static. However, the dynamic system allows vehicles to be driven, which can save time.

Both systems can be integrated into one system and solve the issues related to the limited capacity of rechargeable batteries. According to Bombardier (Pérez R et al. 2016), with dynamic charging, an electric bus can use only 1:5 of the batteries capacity compared to a bus with a static charging system. This leads to a bigger space for the passengers, 25% more compared to the classic electric bus. The dynamic charging system also benefits battery life as they claimed it could be improved to a 60% extension.

2.4.3 Pavement Design to Embed DPT System

There are few basic requirements regarding the DPT/DWPT system, including the transmitter coils of the system need to be embedded close to the surface to increase the charging efficiency. Complex cables need to be enclosed inside the pavement. The wireless charging system needs to be protected by a certain type of materials. Non-ferromagnetic material is preferred since the charging performance will be reduced when it is enclosed or surrounded by ferromagnetic material (such as steel) as there will be eddy current losses. Referring to [21], the DWPT system for RPEV is covered with sand/epoxy/fiberglass filler. Concrete material is used for coverage, but it was found that concrete, which is lower in cost, could not provide enough protection for the cores as sand/epoxy/fiberglass. This system is laid on the concrete pavement. Then the entire roadway was covered with a slurry seal with a thickness of fewer than 5.1 mm.

One of the first developments of the DWPT system, including pavement design for the electrified roadway, was done in the PATH program. The system utilized a roadway inductor, which consists of the following basic components: the inductor modules and the conductor (as seen in **Error! Reference source not found**).

These components were designed to withstand full traffic loads and require little maintenance.

Another concept of embedded IPT components on the prefabricated slab was patented by PRIMOVE. The 25 cm thick slab lay on an 8 cm bituminous base with aluminum supports. The bond between the slab and the bituminous layer is strengthened with 2 cm of grouting. Slabs were tested on the accelerated pavement testing facility of IFSTTAR in Nantes, France. It is an outdoor facility with a circular carousel equipped with four 20 m long arms and has a mean radius of 17.5 m and a width of 6 m, thus a total length of 110 m. The system, illustrated in Fig. 2.6, is claimed to be durable, and it has the performance of holding one million of 6.5 tons (traffic) load cycles. This is equivalent to traffic of 150 buses per day for 20 years. No particular surface treatment was reported [18].

Another example is given by eRoadArlanda, where they had installed the first charging rail on a public road with a length of 2 km in June 2013. The introduced system is railings based embedded in the pavement. They explained that it is critical to combine battery power with direct power feeds while the vehicle is on the move. On minor roads, the vehicles still can run on batteries. But on the major and frequently used roads, the batteries will be recharged continuously. Therefore, the battery-powered vehicles would only need to be designed to be driven with a section of road with an electrical feed. EV can also be recharged at fixed stations. This allows optimum function for the EV since long-distance driving will be not necessary until the EV enters a major road with dynamic charging. There is only a need for a grid with a size of about 45 km in length on approximately 20.000 km of Sweden's highways with this system. This grid will result in smaller batteries for EV, which means lower cost and a better environment [6].

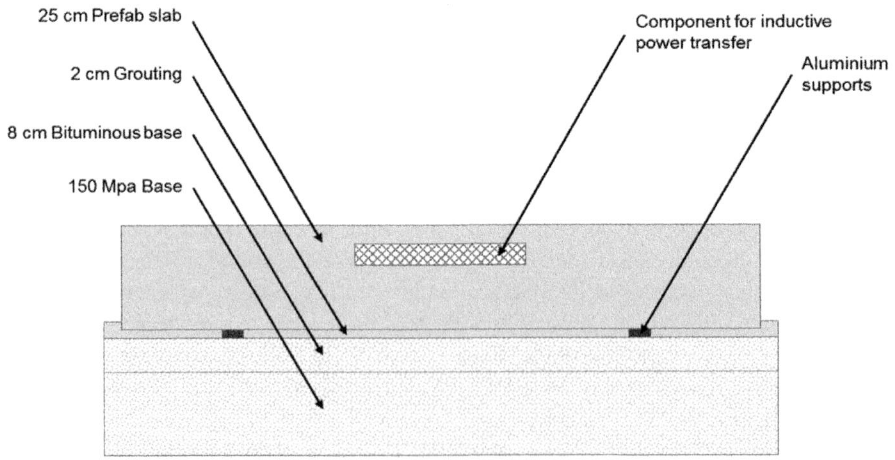

Fig. 2.6 Cross section of PRIMOVE prefabricated slab [17]

References

1. Bombardier (2015) Datasheet primove E-Bus
2. Brecher A, Arthur D (2014) Review and evaluation of Wireless Power Transfer (WPT) for electric transit applications
3. Bresnock A, Miller M, Lechner E, Shladover S (1993) Roadway electrification: regional impact assessment. Trans Res Rec 1416
4. Choi SY, Jeong SY, Gu BW, Lim GC, Rim CT (2015) Ultraslim S-type power supply rails for roadway-powered electric vehicles. IEEE Trans Power Electron 30(11):6456–6468. https://doi.org/10.1109/TPEL.2015.2444894
5. Department of Statistics Singapore (2017) Singapore in Figures 2016, Singapore
6. eRoadArlanda (2018) Roads of the future—electrified roads: vehicles are recharged while driven, Sweden
7. GovTech (2017) Public transport utilisation: average daily public transport ridership. Singapore Open Data Licence. http://www.mytransport.sg/content/mytransport/home/dataMall.html#Facts_&_Figures. Accessed 12 Apr 2017
8. Hernando D, del Val MA (2016) Guidelines for the design of semi-rigid long-life pavements. Int J Pavement Res Technol 9(9):121–127. https://doi.org/10.1016/j.ijprt.2016.03.003
9. Highways Agency (2014) Preparing the strategic road network for electric vehicle
10. Hutin M, Leblanc M (1894) Transformer system for electric railways
11. IPT Technology (2016) Competitive, clean, and efficient public transport with IPT charge bus. Efringen-Kirchen
12. Jung YS, Zollinger DG, Tayabji, SD (2007) Best practices of concrete pavement transition design and construction. FHWA Report No. FHWA/TX-07/0-5320-1
13. Land Transport Authority (2008) Code of practice for works on public streets, 6th edn. Land Transport Authority, Singapore
14. Land Transport Authority (2013) Land transport master plan 2013, Singapore
15. Land Transport Authority (2014) Singapore land transport statistics in Brief 2013, Singapore
16. Mark H (2017) Electric cars may stall without a battery revolution. https://www.theguardian.com/sustainable-business/2017/jul/19/electric-cars-battery-revolution-tesla-volvo
17. NCSS (2017) Singapore emission profile. https://www.nccs.gov.sg/climate-change-and-singapore/national-circumstances/singapore's-emissions-profile. Accessed Jan 2018
18. Nguyen ML, Hornych P, Karzreho JP, Perez S (2014) Accelerated full scale test on prefabricated concrete slab pavement for electrical supply by induction of urban transport sytems. In: 12th International symposium on concrete roads, Prague, Prague
19. Pérez RSA, Nguyen, ML, Hornych P, Curran E (2016) Implementing recharging inductive technology on heavy duty pavement bringing unlimited autonomy to electrical vehicles. In: 8th RILEM international conference of mechanisms of cracking and debonding in pavements
20. Siau ME (2018) Singapore cut carbon intensity. https://www.todayonline.com/singapore/singapore-cut-carbon-intensity-30-cent
21. System Control Technology (1994) Roadway powered electric vehicle Project track construction and testing program phase 3D
22. The Economist (2017) An infrastructure for charging electric vehicles takes shape. https://www.economist.com/business/2017/09/07/an-infrastructure-for-charging-electric-vehicles-takes-shape
23. Transport Research Laboratory (2015) Feasibility study: powering electric vehicles on England's major roads, Birmingham
24. TUMCREATE (2018) The center for electromobility in megacities. https://www.tum-create.edu.sg/about/center-electromobility-megacities. Accessed 2017
25. Wu DQ, Zhang Y (2011) The semi-rigid pavement with higher performances for roads and parking aprons. CAFEO 29, sustainable urbanization—engineering challenges and opportunities, Brunei Darussalam, 27–30 Nov 2011

Chapter 3
State of the Art: Engineered Cementitious Composites Precast Ultra-Thin Whitetopping (ECC-PUTW)

In this chapter, the design concept of Precast Ultra-Thin Whitetopping (PUTW) is reviewed. PUTW concept design offers an option for modification so it can be utilized as electrified roadways. PUTW uses different material than rigid or flexible pavement to avoid the usage of steel reinforcement. Engineered Cementitious Composite (ECC) was then proposed for PUTW development. ECC material is also discussed in this chapter.

PUTW pavement system was developed based on Ultra-Thin Whitetopping (UTW) system with prefabricated slab method to reduce the construction time, which will cause high user delay cost. UTW or thin bonded concrete overlay over asphalt is commonly known as one good solution for preventing rutting due to heavy loads. At the same time, Precast Concrete Pavement (PCP) offers a reduction of the road closure time while providing the high performance of a conventional rigid pavement. However, joint and load transfer devices in PCP are rather complicated. In PUTW, normal joints with dowel and tie bars are not being used. The bond between slabs highly depends on the bond performance between PUTW and the asphalt layer. PUTW can be installed on milled asphalt or newly paved asphalt. However, good-performance asphalt is required.

3.1 PUTW Design

PUTW pavement system is characterized by the material properties, interface bonding in the composite layer, slabs dimensions and joint performance. PUTW is designed as a high-performance pavement, which is suitable for heavy traffic junctions and bus stops. A material called Engineered Cementitious Composite (ECC) is used.

ECC is a unique type of high-performance fiber-reinforced cementitious composite. The most notable characteristic of ECC, which is achieved by the formation of multiple microcracks is high tensile ductility. The compressive strength of

A. A. Bawono, *Engineered Cementitious Composites for Electrified Roadway in Megacities*, Springer Theses, https://doi.org/10.1007/978-3-030-88542-7_3

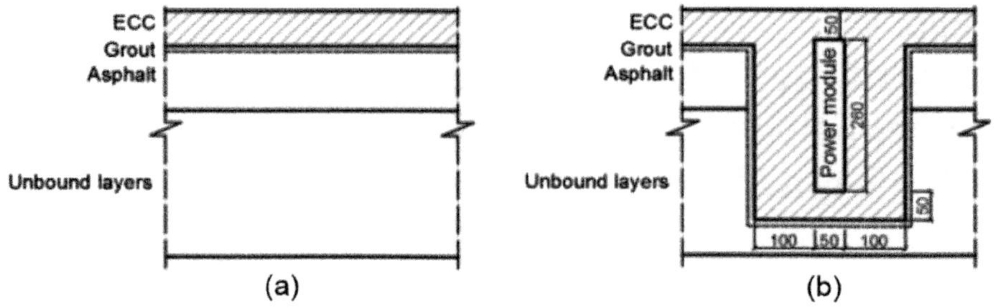

Fig. 3.1 Design of **a** PUTW **b** electrified PUTW [22]

ECC is similar to normal concrete, which usually is in the range from 30–90 MPa. PUTW uses a very thin prefabricated slab. The slab assuredly has to be able to withstand the traffic and construction loads (example given is slab lifting process). Due to its material properties, using reinforcement mesh is no longer needed. On the other hand, a metallic reinforcement mesh is not a good option for electrified roadways as it will interfere with the electromagnetic field. Therefore, ECC material is chosen as a potential candidate for PUTW (Fig. 3.1).

PUTW is designed to keep the asphaltic wearing course layer to a certain thickness which a PUTW slab will replace. Assuming the rest of the asphaltic layer still has good performance, with an even thin (50 mm) PUTW slab, the system can handle 25% more traffic load than conventional rigid pavement [22].

PUTW design can be modified to accommodate electrified roadways, which uses a power module. The structural design is similar to the normal PUTW concept. However, the modified PUTW design is housed by a culvert to protect the power module. When the S-Type power supply module of the latest KAIST OLEV system (5G) is proposed to be embedded in PUTW, the top and bottom covers of the culvert need to have a thickness of 50 mm.

In PUTW, concrete and asphalt layer are bonded together by grouting to form a composite layer. The good bonding strength is the foundation for the durable and long-term performance of the system. As there is no dowel or tie bar to handle the load transfer, the bonding that keeps the positions of PUTW slabs is fulfilled by the asphalt layer. If there is no bonding, the PUTW slab will drift and might endanger road users. Moreover, similar to UTW, the bonding is particularly important for the load transfer between slabs [3, 10, 22, 24]. The bond strength requirement is fulfilled due to the help of grout material between the slab and asphaltic layer and the convex nodes on the bottom surface [22].

PUTW slab needs to be lifted, whether during the manufacture or installation process. During the manufacture, the slab needs to be lifted from the formwork to the storage place. Then the slab will be lifted from the storage place to the lorry to be transported to the site. Finally, the slab is then lifted from the lorry to be placed in the designed area. The slab needs to be levelled during the installation process according to the standards as the roadway geometries are ideally designed with required crossfall and slope. This can be done with the help of lifting and levelling device. The device is using one plate with a different type of bolts. The lifting bolt is normally equipped with a hook to lift the slab, while the levelling bolt is fitted with a metal lug head, so it can be rotated to set the level [23].

PUTW requires a higher initial construction cost compared to conventional concrete- or flexible pavements. However, with the advantage of the prefabricated slab method, PUTW offers lower user delay cost due to minimum delay road closure. Additionally, less pavement rehabilitation and less maintenance result in a lower life cycle cost for the PUTW system. The previous study compares the Life Cycle Cost Analysis (LCCA) between the two pavements: cast-in-place JRCP and PUTW. The results showed that the initial cost for PUTW is relatively high, where user cost is the most significant component in the LCCA of cast-in-place JRCP. However, due to rapid installation, the user cost of PUTW construction is far less than the cast-in-place JRCP. The difference of LCC is up to 2.15 times when the traffic diversion ratio is 0. However, when the traffic diversion ratio is 30%, the LCC of the two alternatives are quite close to balance. In conclusion, PUTW has the advantage of economical LCCA, depending on construction site location and traffic volume.

3.2 Engineered Cementitious Composite (ECC)

3.2.1 The Application of ECC

Engineered cementitious composites (ECC) is a unique class of the high-performance fiber-reinforced cement-based material showing extreme tensile ductility of several percent [4, 25]. ECC has been used in various applications, such as to make coupling beams in high-rise buildings [14], to create a bridge deck link slab replacing expansion joints in a bridge [9], and to retrofit the surface of irrigation channel [28], dam [29], and earth retaining wall [8]. A few studies have reported the potential use of ECC for rigid pavement applications. On top of that, the reflective cracking failure, which commonly occurred in overlay, can be eliminated because the ECC's high ductility overlay is able to trap the kinked-out microcracks from the interface [40]. Compared to conventional concrete and hot mix asphalt (HMA) overlay systems, the ECC overlay offers lower life cycle cost over 40 years of service life that reduces total life cycle energy by 14%, Greenhouse gas (GH) emissions by 43%, and costs by 40% [41]. It is reported that ECC can be used for repair material to provide better system performance as it is stronger, more ductile, more energy absorbing, and it shows better crack control compared to concrete [17]. A similar scenario has been reported by Qian and Li [25], that ECC pavement will be more favorable for medium to heavy traffic due to the efficiency of pavement thickness, jointless/larger joint spacing, and less maintenance required. Recently, ECC is applied on an 80 m jointless test track. And it shows that this material can be produced in large quantity and at high quality. The long-term monitoring showed that the material could handle high traffic loads, and the abrasion was low [20].

3.2.2 Micromechanics-Based Design of ECC

The development of ECC design is ideally based on the performance-driven design approach (PDDA) (as can be seen in Appendix A), which emphasizes micromechanics models as the unifying bridge between mechanical properties and material microstructure. With PDDA, the imposed requirements of structural performance, e.g., deformation capacity, ductility, and load capacity, are depicted into mechanical properties at the material level, e.g., strength, ultimate strain, toughness, and maximum crack width. A great advantage is PDDA offers systematic engineering solution rather than trial-and-error empirical experiments [15].

To achieve multiple cracking and tensile strain-hardening behavior in ECC, there are two conditions need to be fulfilled:

- Crack tip toughness J_{tip} has to be less than the complementary energy J_b', which is calculated from the bridging stress σ with crack opening δ curve.

$$J_{tip} \leq \sigma_0\delta_0 - \int_0^{\delta_0} \sigma(\delta)d\delta \equiv J_b' \tag{3.1}$$

where

$$J_{tip} = \frac{K_m^2}{E_c} \tag{3.2}$$

- The tensile first crack strength σ_{fc} must not exceed the maximum bridging stress σ_0.

$$\sigma_{fc} < \sigma_0$$

According to Li et al. [15], Eq. (3.1) implies the stress-crack opening relationship $\sigma(\delta)$, which is regarded as a constitutive law of fiber reinforced composites, is derived from analytic analysis of fracture mechanics, micromechanics, and statistics. Where σ_0 is the maximum bridging stress corresponding to the crack opening δ_0, while E_c is the composite elastic modulus.

Due to a distribution limitation of pre-existing flaw size, a sufficient margin between complementary energy J_b' and crack tip toughness J_{tip} is desired for developing saturated multiple cracking. Examples of methods to achieve it are given:

- Reducing J_{tip}, which is correlated to matrix toughness, or
- increasing J_b', which is more reasonable because low matrix toughness implies low first crack strength.

Table 3.1 Parameters in PDDA approach for ECC design

Fiber		Matrix		Interface	
V_f	Volume fraction	K_m	Fracture toughness	τ_o	Interface frictional bond
l_f	Length	E_m	Elastic modulus	G_D	Interface chemical bond
d_f	Diameter	a_0	Initial flaw size distribution	f	Snubbing coefficient
E_f	Elastic modulus	–	–	f'	Strength reduction factor
σ_{fu}	Tensile strength	–	–	β	Slip-hardening coefficient

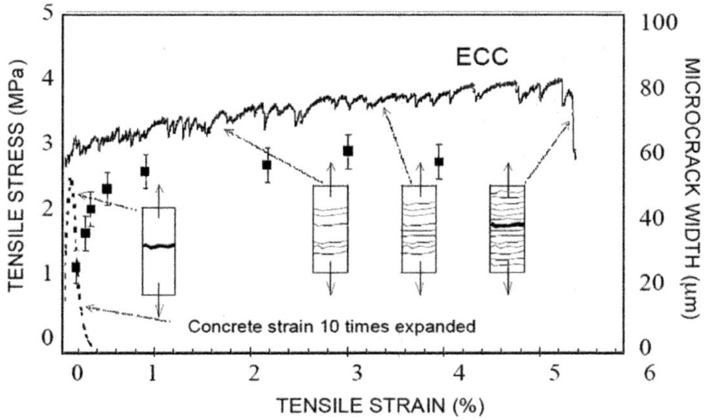

Fig. 3.2 Strain-hardening behavior of ECC material [13]

In the microstructure tailoring part, three sets of parameters are considered, including fiber, matrix, and interface (fiber-matrix interaction) phase. Parameters are summarized in Table 3.1 (Fig. 3.2).

3.2.3 Ingredients Used in ECC

ECC consists of various ingredients with different specification. In this study, different mixture designs with different material were tested. The density of each material is summarized in Tables 3.2 and 3.3.

In this study, Type I Portland cement (CEM I 52.N) is used. Another type of cement used in ECC is Type I white cement. Type I white cement is typically used

Table 3.2 Density of materials for ECC

Material	Cement/ White Cement	Micro-Silica Sand	Water	Fly Ash	GGBS	Fiber PVA	Fiber PE	Super -plasticizer
Density (g/cm^3)	3.15	2.65	1.00	2.20	2.91	1.30	0.96	0.30

Table 3.3 Density of suitable aggregate for skid resistance improvement

Material	Corundum	Silica Sand	Granite	Steel Slag
Density (g/cm^3)	3.95	2.65	2.60	3.20

for ECC for architectural applications [11, 38]. Micro-silica sand with particle size: 0.08–0.12 mm is used in ECC for this study. Normal tap water is used for ECC.

Fly ash is commonly used to reduce the chemical bonding while increasing frictional bonding of the fiber and matrix interface. Furthermore, fly ash replacement of cement means to lower the matrix toughness. A matrix with higher fly ash content can compensate for the material components' rate sensitivity that limits the tensile ductility at higher loading rates [36]. Thus, fly ash improves mixture workability and material sustainability. On the other hand, high content fly ash leads to slower strength development at the early stage. Similar result with Yang and Li's test, incorporating fly ash into matrix tends to reduce both the interface bond and matrix toughness [33]. Fly ash class F with particle size: 0.01–0.15 mm is used in the ECC mixture for this study.

The use of fiber to maintain the strain hardening is understandably kept to the minimum as it is more expensive than the other materials used in ECC. Using fiber content in a low volume will lead to normal FRC tension-softening behavior. On the other hand, using fiber content greatly value leads to workability issues [12]. In this study, two different types of fibers are used. PE fiber is produced by Honeywell with Spectra® 1000 type, with a diameter of 23 μm, a length of 12 mm and 19 mm. While PVA fiber is produced by Kuraray REC(S) 15 type, with a diameter of 40 μm, length of 12 mm, and 1.2% surface oiling content.

PVA is an organic fiber that offers advantages of high tenacity and modulus of elasticity when compared with other general organic fiber. PVA also provides strong bonding strength between fiber and cement matrix, and it is durable [31]. PVA fibers are hydrophilic, which makes them strongly bond to the matrix. This leads to limit the multiple cracking effects and lower strain hardening behavior for the composite. Reducing the chemical de-bonding energy enhances the complementary energy by minimizing premature fiber breakage during the matrix interface de-bonding process [26]. PE fiber has an extremely high melt viscosity, toughness, cut and wear resistance, and very good chemical resistance. PE fibers enhance the toughness and flexural strength of ECC [30]. Nevertheless, the application of PE fiber in cement composite is limited due to the high cost [31]. In a single fiber pullout test, PE fiber has shown the increase of the interface friction during fiber slippage induced a slip hardening behavior, leading to the increase of strain hardening for the composite. However, the slip hardening should be limited to avoid fiber rupture. This is because the fiber pullout is causing a higher ductility compared to fiber rupture [26]. Composite with PVA gives better results in compressive, split tensile and flexural strengths, followed by PE, PET and PP fiber as reinforcement [31]. PVA fiber is considered as low modulus and low strength fiber resulting in low ultimate strength with high ductility of the composite. While PE fiber, has about 2–3 times higher than the PVA in terms of modulus with tensile strength is similar or to higher than the steel fiber. Therefore, a composite with PE fiber might have high strength and high ductility [1].

As the oiling content of surface coating increases, both frictional stress and interface fracture energy will decrease significantly. As shown in Fig. 3.3, the complementary energy of the fiber bridging stress increases with the oil content given [33]. A surface coating content between 0.8–1.2% by weight of fibers tends to lower then interface chemical and frictional bond properties to a level that causes the critical fiber volume fraction to drop to a minimum of about 2% [16] (Table 3.4).

The usage of superplasticizer is to increase workability characteristics. It allows the material to be produced with relatively low water-cement ratios without experiencing loss in strength. The usage is beneficial as ECC is relatively hard to be levelled due to fiber existence. However, the content of the superplasticizer needs to be controlled. As when the dosage is too much may lead to segregation. In this study, the superplasticizer is produced GCP Applied Technologies with type of ADVA® 181. It is a polycarboxylate-based superplasticizer with no added chloride in liquid form with a specific gravity of 1.10 kg/liter.

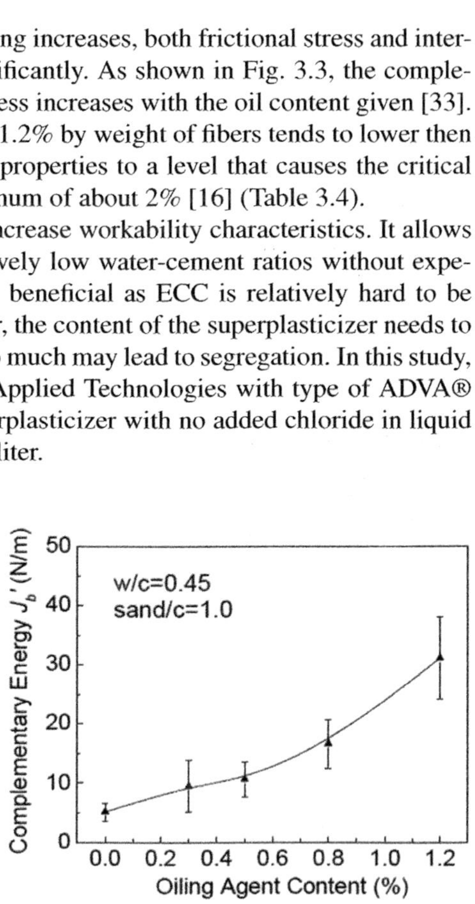

Fig. 3.3 Effect of fiber surface oiling content [16]

Table 3.4 Fiber specification

Material	Length (mm)	Tensile strength (MPa)	Elastic modulus (GPa)	Density (g/cm^3)
PVA fiber	12	1560	40	1.30
PE fiber	12 / 19	3250	113	0.96

Viscosity Modifying Agent (VMA) is used in the form of water-soluble polymers to increase the viscosity of concrete/ECC mix. There are at least three mechanisms of VMA action. The first one is adsorption as results of long polymer chains relate to water molecules adsorbing some part of the mix water and increasing their volume. Secondly, association as in the molecules of adjacent polymer chains exhibit an attractive force by reducing the mobility of water molecules leads to gel form, thus increasing the viscosity. The third is splicing, which occurs at a low shear velocity, increasing the apparent viscosity [5].

In conventional concrete, the Portland cement paste that suffers from autogenous and drying shrinkage makes up less than 20% of the total volume [21]. The Portland cement paste in ECC makes up moreover 50% of the total volume [37]. It indicates that ECC would experience greater shrinkage, and when constrained, it could lead to shrinkage cracking. In a pioneer project in Michigan, U.S., shrinkage cracks were observed on the bridge link-slabs made of bendable concrete after eight days since paving. The maximum crack width can be 0.2 mm [39]. In the current study, a shrinkage reduction agent in powder form (Denka Power CSA Type-S) was added in the standard mix proportion design of bendable concrete. It has a specific gravity of 2.95–3.15 g/cm^3 with Blaine fineness of more than 2000 g/cm^2.

A specific type of aggregate can be utilized to provide high skid resistance. However, there is less research working on the modification of ECC mixture design for pavement application. As can be seen below, different types of aggregate have different performance and affect the mechanical properties of concrete/ECC differently (Table 3.5).

Corundum is a crystalline form of aluminum oxide (Al_2O_3). Corundum is extremely hard. It is rated at nine on Mohs scale of mineral hardness (diamond rated as 10, while the most common mineral in aggregates, quartz, only rated as 7) that can scratch almost every other mineral [32]. Application of surface treatment using corundum has been reported. According to the report of Kreisverkehrsflächen in Beton: Erfahrungen in der Schweiz, Deutschland und Österreich [34], corundum material with the sizes of 1–3 mm has been applied to provide a high grip of the road

Table 3.5 Suitable aggregates for high skid resistance

Type of aggregate	Aggregate size	Product
Corundum	1–3 mm	–
Bauxite (China-Guyana) Granite	1–3 mm 1–3 mm	Crafco HFS
Basalt and Granite Silica Sand	2–3 mm 2–3 mm	Flexogrid, modified EP-5 overlay
Steel slag	3–4 mm	Italgrip
Limestone dolomitic	1–3 mm	Safelane (Cargill)
Granite	1.5–3.2 mm	Safe-T-Grip
Calcined bauxite	1–3 mm	Tyregrip

Table 3.6 Chemical specification of corundum

Chemical specification (%)						
S_iO_2	Al_2O_3	Fe_2O_3	T_iO_2	CaO	MgO	FeO
14.61	61.80	12.25	8.14	N.A	N.A	N.A

surface by spraying evaporative protection and followed by covering an area with thermos-mats (Table 3.6).

Calcined bauxite is produced by heating bauxite to 1000–1200 °C, which increasing the alumina content. Calcination can occur either in nature during periods of metamorphism or can be processed artificially in furnaces. The bauxite itself is a material rich in hydrated aluminum oxide [32]. Many high skid resistance surface treatments use calcined bauxite aggregate successfully. One of the examples is the Tyregrip product. Washington Department of Transport (WSDOT) applied TyreGrip on the ramp carrying southbound traffic from Southeast 164th Avenue to State Road 1st, which has a downhill before entering the sharp horizontal curve of the ramp that merging onto State Road 14. Increasing the frictional properties of the surface pavement was proposed to eliminate the accidents occurring in this section [2].

Granites are commonly composed of potassium feldspar, plagioclase feldspar, and quartz. In addition, this material contains a high concentrate of silica (SiO_2) and alumina (Al_2O_3). Due to its mineral composition and interlocking matrix, this material is hard and has an abrasive resistance. The toughness of granite is usually superior to sandstone, limestone or marble. Granites normally can provide a PSV of 62 or greater [6].

Silica material has high abrasion resistance and high thermal stability. It can provide high skid resistance for surface treatment. Silica occurs in nature sandstone, silica sand or quartz. Silica can also exist in an amorphous form or crystalline forms [27]. The FlexoGrid system uses silica to provide surface friction. The application is designed to be installed on conventional concrete, concrete-filled steel, grid decks, steel plate decks, fiber reinforced polymer decks, and asphalt. However, based on the evaluation of the performance of the FlexoGrid system on a highway bridge in the state of Iowa, the friction measurement only recorded 67.5 SN after the application and then decreased to 64.5 SN after four years.

Slag is produced during the separation of the molten steel from impurities in steel making. It must be crushed and screened to produce a suitable aggregate for surface treatments [27]. Steel slag is used in the ItalGrip system. It is an ultra-thin polymer concrete that provides the skid resistance of pavement surface. It consists of a two-part polymer resin that is sprayed on the pavement surface and covered with a steel slag with the size of 3–4 mm. The Italgrip system can be placed on either HMA or concrete pavements [7]. LTA of Singapore also used steel slags for their surface treatment aggregates since this material can be easily found in the local area and also can provide good abrasive.

Limestone is categorized as sedimentation rock that consists of more than 50% carbonate. It is made of the mineral calcite from the beds of evaporated seas, lakes, and sea animal shells [19]. Limestone dolomitic aggregates have been used as the main material for the SafeLane product. With the aggregate median size of 4 mm and gradation from 3/8 inch to sieve number 30, the overlay has been done with the help of two-layer epoxy. SafeLane product can improve the skid resistance to 0.50 to 0.57 of Griptester number [27].

3.2.4 Mixture Design of ECC

A typical mixture design of ECC is called ECC-M45 (see Table 3.7), has the largest experimental dataset [13]. ECC incorporates fine silica sand with sand to binder ratio of 0.36 to maintain sufficient stiffness and volume stability. The silica sand has a maximum grain size of 250 μm. ECC-M45 uses fly ash (Type F), high range water reducer (HRWR), and PVA fiber with a length of 12 mm and 39 μm in diameter. The PVA fiber is surface coated by a proprietary oiling agent (1.2% by weight) to reduce the fiber/matrix interfacial bonding.

It was proposed for PUTW to use modified ECC with local materials in Singapore: ground granulated blast-furnace slag (GGBS) and river sands instead of fly ash and micro-silica sand can be used. The raw materials of this modified ECC include Type I OPC (CEM I 42.5), GGBS provided by Engro Co. Ltd, river sands sieved to below 600 μm in particle size, and PVA fibers from Kuraray Co. Ltd. The fibers are 12 mm in length and 39 μm in diameter. The nominal fiber strength is 1600 MPa. Superplasticizer (SP) was also included to ensure the workability of ECC. Three different mix designs are shown in Table 3.8.

Table 3.7 Mix design of ECC (M45) [13]

OPC	Fly Ash	Sand	Water	HRWR	Fiber (vol.%)
1.0	1.2	0.8	0.56	0.012	2

Table 3.8 Mix design of modified ECC-GGBS [22]

ECC-	OPC (kg/m^3)	GGBS (kg/m^3)	Sand (kg/m^3)	Water (kg/m^3)	PVA fiber (kg/m^3)	SP (liter/m^3)	GGBS/ binder ratio (%)
GGBS0	1411	0	282	423	26	3.6	0
GGBS30	976	418	279	418	26	3.0	3
GGBS60	554	827	276	414	26	2.4	60

Table 3.9 Typical mechanical properties of ECC

Compressive strength (MPa)	First crack strength (MPa)	Ultimate tensile strength (MPa)	Ultimate tensile strain (%)	Young's modulus (GPa)	Flexural strength (MPa)
20–95	3–7	4–12	1–8	18–34	10–30

3.2.5 The Performance of ECC

The most notable characteristic of the ECC is the high tensile ductility, which is represented by a uniaxial tensile stress–strain curve with strain capacity up to 5% due to the performance of multiple micro-cracks formations. The formation of multiple micro-cracking is the reason how ECC can behave similarly to metal. Typical mechanical properties for ECC is summarized in Table 3.9 [15]. The compressive strength of ECC ranges from 20–95 MPa with an elastic modulus range between 18–34 GPa. The flexural strength (modulus of rupture) of the ECC material can be achieved quite high up to 10–30 MPa. It has also been reported that ECC has lower water permeability and lower effective chloride diffusivity in the presence of micro-cracks [9].

The modified ECC-GGBS have shown good performance. Test related to mechanical properties only performed on flexural strength with four points bending tensile test. It is reported that the average value of ECC-Cor is 10.50 MPa with 7.00 MPa and 15.20 MPa as maximum and minimum strength, respectively [22].

3.3 PUTW Slab Manufacture

The process of PUTW slab manufacture consists of two main steps, including (1) preparation of the formwork and (2) manufacture process of PUTW slab. The first step, manufacturing the PUTW slab, starts with the preparation of steel formwork. In this study, the steel formwork was prepared with an inner dimension of 3400 mm × 2400 mm and a thickness of 70 mm. The formwork is equipped with removable steel walls so the walls can be removed after the slab is hardened. Detailed dimension and location of the grouting port as well as the lifting and levelling devices, can be seen in Appendix A.

A Rubber mat with convex nodes was placed inside the formwork. Six PVC pipes were prepared as grouting ports embedded in the PUTW slab. Additionally, four lifting and levelling devices were also prepared to be embedded in the PUTW slab. All of the grouting ports and levelling device need to be temporarily closed with tape to avoid the penetration of ECC material during the casting process. The tape can be only opened when the ECC material is hardened.

The next step is to cast the ECC material into the formwork. Two different ECC materials are used. The first layer is made of ECC with a thickness of 60 mm. Then the second layer made of modified ECC (with coarse aggregates) with a thickness of 10 mm is cast on top of the first layer. Both layers are cast fresh on fresh to achieve homogenous material, which means the second layer should be cast within one hour. Otherwise, the first layer of ECC will get hardened. Yet, during the casting, it was found that levelling ECC is relatively difficult as the fibers are well embedded into the matrix.

The curing process is then started. The slab is then brushed after 18–36 h with the brushing machine to expose the coarse aggregates. The brushing process cannot be directly done after the casting process as the ECC is not hardened yet. But it should be not delayed too long as when the ECC is fully hardened. It will be very difficult to remove the cement paste resulting in fewer coarse aggregates appear on the surface.

3.4 PUTW Construction

The PUTW construction consists of different steps: lane closures, asphalt pavement preparation, slabs installation, slabs levelling, and grouting process. The construction process will take 4–8 h [35] (Fig. 3.4).

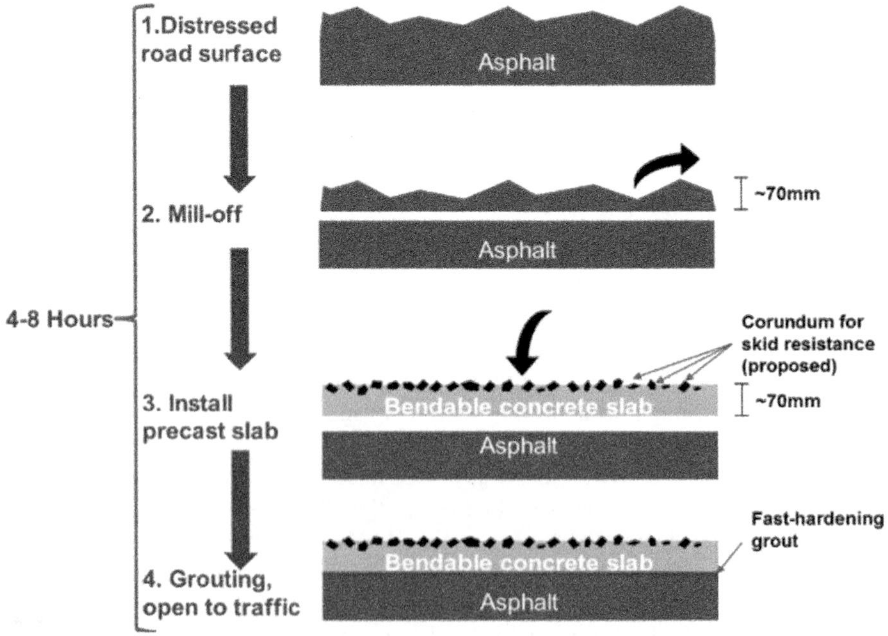

Fig. 3.4 PUTW construction process

Before the construction begins, the site needs to be closed. In this study, Singapore standards are followed. The work zone requires at least two road lanes. The rest of the road lanes are still opened for traffic. The lane closure must comply with the authority's (LTA's) work zone traffic control guidelines (LTA [18]. Depending on the purpose, some temporary facilities (e.g., bus stop) will be provided together with temporary guidance signs.

Since asphalt acts as the support layers for the PUTW slab, it is decisive for the overall performance of the PUTW pavement. Therefore, except for minor surface distress, the asphalt pavement structure must be in good condition. The asphalt layer needs to be milled with a minimum thickness of 70 mm. Then the milled asphalt surface will be cleaned using a brush and air compressor.

By using a lifting device, the PUTW slab is then lifted and installed within the designed area. The slab is then levelled with the levelling device to meet the cross fall and slope criteria. Once the slab is levelled, the fast-hardening grout is then injected through the port. It is important to reduce the amount of air bubbles trapped between the slab and the asphalt layer underneath. Therefore, it is required to fill in the grout until the excess of grout is flowing out through the other ports. It is also required to check the gaps between the slabs. Ideally, all the gaps should be filled with grout.

References

1. Ahmed SFU, Ronnie Z, Abdullah M, Rahim SZA, Suandi MEM, Saad MNM, Ghazali MF (2017) Ductile behavior of polyethylene fibre reinforced geopolymer composite. MATEC Web Conf. 97(1):1047. https://doi.org/10.1051/matecconf/20179701047
2. Anderson KW, Russel M, Sheets J, Burkey D, Uhlmeyer JS, Weston J (2012) Evaluation of tyregrip® high-friction surfacing, Washington
3. Eid JA (2012) Theoretische und experimentelle Untersuchungen dünner Betondecken auf Asphalt
4. Fischer G, Li VC (2003) Deformation behavior of fiber-reinforced polymer reinforced engineered cementitious composite (ECC) flexural members under reversed cyclic loading conditions. ACI Struct J 100(1):25–35
5. Grabiec AM (2013) Influence of viscosity modifying agent on some rheological properties, segregation resistance and compressive strength of self-compacting concrete. J Civ Eng Manag 19(1):1–8. https://doi.org/10.3846/13923730.2012.737830
6. Huhta R, Meyer D, Zelnak P (2001) The aggregate handbook
7. Kuemmel DA, Jaeckel JR, Satanovsky A (2000) Investigative study of the Italgrip system: noise analysis. Milwaukee, Wisconsin
8. Kunieda M, Rokugo K (2006) Recent progress on HPFRCC in Japan required performance and applications. J Adv Concr Technol 4(1):19–33. https://doi.org/10.3151/jact.4.19
9. Lepech MD, Li VC (2009) Water permeability of engineered cementitious composites. Cement Concr Compos 31(10):744–753. https://doi.org/10.1016/j.cemconcomp.2009.07.002
10. Li Z, Vandenbossche JM (2013) Redefining the failure mode for Thin and Ultra-thin Whitetopping with a 1.8-m x 1.8-m (6- ft x 6-ft) joint spacing. TRB 2013 Annual Meeting, Washington, DC
11. Li M, Li VC (2013) Rheology, fiber dispersion, and robust properties of engineered cementitious composites. Mater Struct 46(3):405–420. https://doi.org/10.1617/s11527-012-9909-z
12. Li VC (2003) On engineered cementitious composites (ECC): a review of the material and its applications. Advanced Concrete Technology 1(3):215–230

13. Li VC (2007) Concrete construction engineering handbook: engineered cementitious composites (ECC)—material, structural, and durability performance. CRC Press, Michigan
14. Li VC, Kanda T (1998) Engineered cementitious composites for structural applications. ASCE J Mater Civil Eng 10(2):66–69
15. Li VC, Wang S, Wu C (2001) Tensile strain-hardening behavior of polyvinyl alcohol engineered cementitious composite (PVA-ECC). ACI Mater J 98(6):483–492
16. Li, VC, Wu C, Wang S, Ogawa A, Saito T (2002) Interface tailoring for strain-hardening polyvinyl alcohol engineered cementitious composite (PVA-ECC). 99(5):463–472
17. Lim YM, Li VC (1997) Durable repair of aged infrastructures using trapping mechanism of engineered cementitious composites. Cem Concr Compos 19(4):373–385
18. LTA Singapore (2018) Code of practice traffic control at work zone. LTA Singapore, Singapore
19. Mohamad M (2005) The use of limestone aggregate in concrete. Malaysia
20. Mueller S, Mechtcherine V (2017) Use of strain-hardening cement-based composites in real scale applications. In: 4th International RILEM conference on strain hardening cement-based composites, Dresden, Germany, 18–20 Sep 2017
21. Neville AM (2012) Properties of concrete. Prentice Hall
22. NguyenDinh N (2017) Precast Ultra-Thin Whitetopping (PUTW) in Singapore. Doctoral Thesis, Technical University of Munich
23. NguyenDinh N, Aguilar OACA, Bawono AA, Lechner B (2017) Connection assembly, levelling or lifting system, and method of levelling slab WO/2020/171775, 12 Feb 2020
24. Pereira DDS, Balbo JT, Khazanovich L (2006) Theoretical and field evaluation of interaction between ultra-thin whitetopping and existing asphalt pavement. Int J Pavement Eng 7(4):251–260. https://doi.org/10.1080/10298430600798895
25. Qian S, Li V (2008) Durable pavement with ECC. Nanjing, China
26. Redon C, Li VC, Wu C, Hoshiro H, Saito T, Ogawa A (2001) Measuring and modifying interface properties of PVA fibers in ECC matrix. J Mater Civ Eng 13(6):399–406
27. Roa J (2008) Evaluation of IFI and high-friction surfaces. Master Thesis, Virginia Polytechnic Institute
28. Rokugo K (2017) Applications of SHCC in Japan—tools and tips for promoting its use. In: 4th International RILEM conference on strain hardening cement-based composites, Dresden, Germany, 18–20 Sep 2017
29. Rokugo K, Kunieda M, Lim SC (2005) Patching repair with ECC on cracked concrete surface. University of British Columbia, Concrete Material, Vancouver, Canada
30. Said SH, Razak HA (2015) The effect of synthetic polyethylene fiber on the strain hardening behavior of engineered cementitious composite (ECC). Mater Des 86:447–457. https://doi.org/10.1016/j.matdes.2015.07.125
31. Sathishkumar P, Sampathkumar P, Karthik M, Vignesh C (2016) Significance of various fibres on engineered cementitious concrete. Int Sci Eng Technol 4(1):284–290
32. Tubey L, Hosking J (1972) Synthetic aggregates of high resistance to polishing—corundum rich aggregates (TRRL LR467)
33. Wang S, Li VC (2006) Polyvinyl Alcohol fiber reinforced ECC: material design and perfomances. In: International RILEM workshop on HPFRCC in structural applications, Michigan
34. Werner R, Peck M (2010) Kreisverkehrsflächen in Beton: Erfahrungen in der Schweiz. Deutschland und Österreich, Swiss
35. Yang EH (2017) Precast bendable concrete paving slab for speedy road construction. Singapore
36. Yang E-H, Li VC (2012) Tailoring engineered cementitious composites for impact resistance. Cem Concr Res 42(8):1066–1071. https://doi.org/10.1016/j.cemconres.2012.04.006
37. Yang EH, Yang Y, Li VC (2007) Use of high volumes of fly ash to improve ECC mechanical properties and material greenness. ACI Mater J 104–168
38. Yang E-H, Garcez E, Li VC (2012) Development of pigmentable engineered cementitious composites for architectural elements through integrated structures and materials design. Mater Struct 45(3):425–432. https://doi.org/10.1617/s11527-011-9774-1

39. Yang EH, Li VC, Qiu J (2015) Early age cracking in a SHCC bridge deck link slab. Seventh congress on forensic engineering. In: ASCE forensic engineering 7th congress, Miami, USA, 2015
40. Zhang J, Li VC (2002) Monotonic and fatigue performance in bending of fiber-reinforced engineered cementitious composite in overlay system. Cem Concr Res 32(3):415–423. https://doi.org/10.1016/S0008-8846(01)00695-0
41. Zhang H, Keoleian G, Lepech M (2008) An integrated life cycle assessment and life cycle analysis model for pavement overlay systems. In: 1st International symposium on life-cycle civil engineering, Varenna, Lake Como, Italy, 10–14 June 2008

Chapter 4
State of the Art: Functional Performance of Pavement

This chapter summarizes the state of the art of the functional performance of the pavement. The functional performance of pavement is vital as it is directly related to the safety and comfort of driving experiences. The following section will state different parameters on functional performance, which included skid resistance, surface water drainage, tire-pavement noise, durability, and self-cleaning.

AASHTO [1] points out the functional performance of the pavement, which is based on its ability to serve the user that has major indicators of roughness, skid resistance, risk of hydroplaning, appearance, and other safety considerations. The statement is amplified by FHWA [30] that states the importance of pavement safety, which can be affected by pavement surfaces commonly characterized by friction, smoothness, tire-pavement noise, and texture. FHWA also indicates that surface durability and drainage are needed to be considered as there must be a balance between providing the required grip and providing a durable surface with the consideration of (life cycle) cost. Surface drainage is a critical factor in pavement safety. Even on the surface pavement with high skid resistance, the safety level can be significantly reduced when the water does not drain efficiently.

Few skid accidents in Singapore have been reported. Skid accidents can occur on different type of pavements: concrete, asphalt and semi-rigid. They are taking place not only in wet condition but also in dry condition. According to statistic, about 7000 accidents occurred in 2012, 2013, 2014, and 2015 [27]. Mainly, these accidents are related to the driving behavior of the driver. Then other factors are the vehicles and road conditions. As shown in Fig. 4.1, in the early years, the number of accidents related to road conditions is similar to the number related to vehicles. But, interestingly, a few years later, Singapore was able to reduce the accidents caused by road conditions almost by half.

A few essential design criteria in functional performance have to be considered and carefully design in pavement engineering. However, all parameters are not linearly correlated with each other. Henry [42] did a study by conducting surveys to roadway authorities on the important parameters on pavement performance. As a result, durability, skid resistance, and splash and spray are coming as the first three priorities. The

A. A. Bawono, *Engineered Cementitious Composites for Electrified Roadway in Megacities*, Springer Theses, https://doi.org/10.1007/978-3-030-88542-7_4

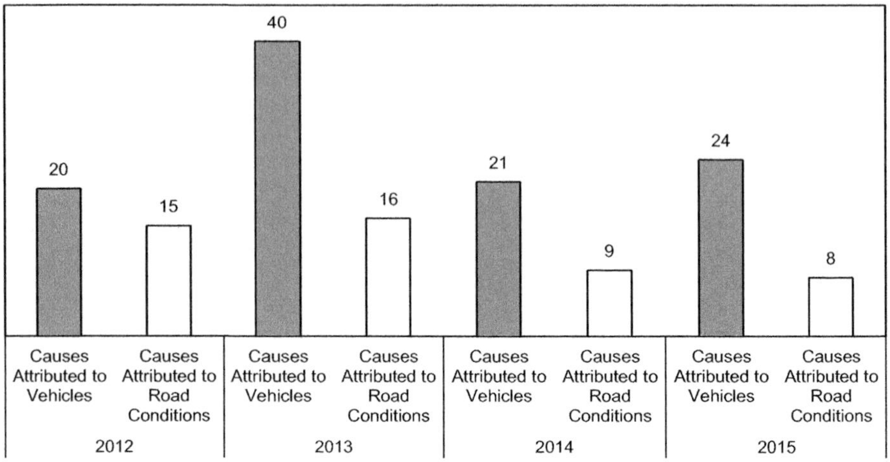

Fig. 4.1 Accidents related to vehicles and road conditions in Singapore

Table 4.1 Importance of pavement based on the functional performance [42]

Design criteria	Ranking	
	United States	Other countries
Durability	1.1	1.3
Skid resistance	1.2	1.4
Splash and spray	2.0	1.8
Exterior noise	2.4	2.2
In-vehicle noise	2.4	2.4
Rolling resistance	2.7	2.7
Tire wear	2.7	2.9

durability is coming as the first prioritization as they are closely related to economic consideration. The performance was rated on a scale of 1–3, where a rating of 1 is very important, and a rating of 3 is relatively unimportant (Table 4.1).

4.1 Skid Resistance

Skid resistance or friction is the counterforce developed due to the movement of the load along the pavement surface that holds the relative movement of a tire to move forward. The generated resistive force is illustrated in Fig. 4.2. The skid resistance of pavement (concrete and asphalt) is essentially affected by the properties of the aggregate, the characteristics of the tire and environmental conditions. The friction coefficient is a non-dimensional coefficient, μ, which is defined as the ratio of the

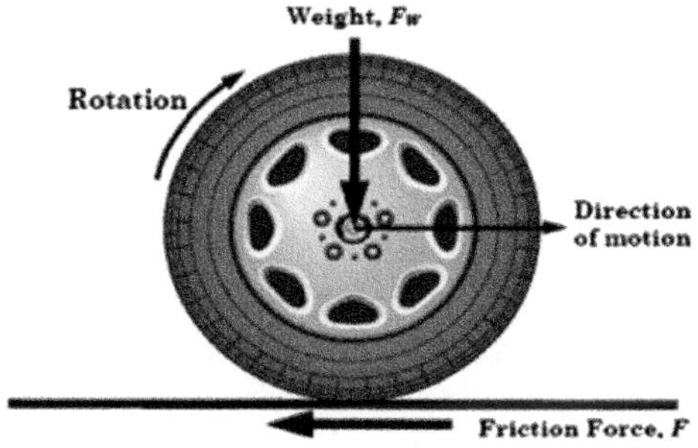

Fig. 4.2 The friction force on tire-pavement

tangential friction force (F) between the tire and the horizontal pavement surface to the perpendicular force or vertical load (F_w).

According to a previous study, skid resistance can be determined in several methods: coefficient of friction (μ), Skid Number (SN), or Friction number (FN) [26]. Meanwhile, in 1992, the World Road Association, Permanent International Association of Road Congresses (PIARC), conducted the experiment to convert results generated by different methods and devices to a common scale, called the International Frictional Index (IFI).

The Friction Number (FN) or Skid Number (SN), which produced by a locked wheel testing device, represents the average coefficient of friction. According to ASTM E274/E274M-11 [10], the skid number or SN is the value to determine the skid resistance of the paved surface from the resulting force or torque record.

$$SN = 100(\mu) = 100\left(\frac{F}{W}\right) \tag{4.1}$$

where

F Tractive force (horizontal force applied to the test tire at the tire-pavement contact patch, N

W Dynamic vertical load on a test wheel, N.

The reporting values range from 0 to 100, with 0 representing no friction and 100 representing complete friction.

FN values are generally designated by the speed at which the test is conducted and by the type of tire used in the test. For example, FN50S = 29 indicates a friction value of 29, as measured at a test speed of 50 miles per hour (81 km per hour) and with a smooth (S) tire. The ribbed type defined as (R).

- Side Force (Mu) Number, also known as sideways force coefficient (SFC), can be obtained by Mu-Meter, which contains two freely rotating test wheels angled to the direction of motion. Based on ASTM E 670-09–Standard Test Method for Testing Side Force Friction on Paved Surfaces Using the Mu-Meter, the relation between MuN and BPN is governed by the equation below:

$$\mathrm{MuN} = 1.017\,\mathrm{BPN} - 20.9 \tag{4.2}$$

where

MuN Mu Number
BPN British Pendulum Number.

- International Friction Index (IFI), which was developed by PIARC, was initiated to standardize the dependency of friction on the tire sliding speed. The IFI consists of two parameters, which are the adjusted value of friction at a certain slip speed (F60) and the Speed Number (Sp).

$$S_p = a + b \times TX \tag{4.3}$$

where

S_p IFI speed number
a, b Calibration constants dependent on the measurement method:
For MPD (ASTM E 1845), a = 14.2 and b = 89.7
For MTD (ASTM E 965), a = −11.6 and b = 113.6
TX Macrotexture measurement, mm.

$$F(60) = A + B \times FR(60) + C \times TX \tag{4.4}$$

where

F(60) IFI friction number
A, B Adjusted value of friction measurement at selected slip speed S to a slip speed of 60 kilometers per hour
C Calibration constants dependent on friction measuring device
FR(60) Calibration constant required for measurements using ribbed tire.

Pavement surface texture influences pavement-tire interactions. Different level of texture influences different effects on pavement-tire interactions. Henry [42] and Sandburg and Ejsmont [73] have studied the effects of various tire-pavement interactions from different ranges of texture wavelengths. Those effects are including friction, interior and exterior noise, splash and spray, rolling resistance, and tire wear. It can be seen from Fig. 4.3 that the skid resistance related to friction is primarily affected by the microtexture and macrotexture.

In terms of the properties of the aggregate, pavement texture affects the skid resistance. Pavement surface texture is defined as the deviations of the pavement surface from a true planar surface. According to the PIARC in 1987, these deviations are divided into three levels of scale, which are defined by the wavelength (λ) and peak-to-peak amplitude (A) of its components. There are three types of texture: microtexture, macrotexture and megatexture [68]. These three levels are defined as follows:

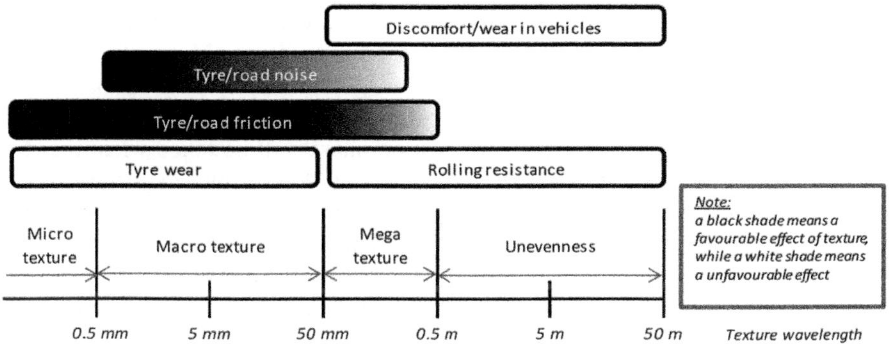

Fig. 4.3 Texture wavelength influence on tire-pavement interactions [68]

- Microtexture ($\lambda < 0.5$ mm, A = 1 to 500 μm). Surface roughness quality at the microscopic level. It is a function of the surface properties of the aggregate particles contained in the asphalt or concrete paving material.
- Macrotexture ($\lambda = 0.5$ to 50 mm, A = 0.1 to 20 mm). Surface roughness quality defined by the mixture properties (shape, size, and gradation of aggregate) of asphalt paving mixtures and the method of finishing/texturing (dragging, tining, grooving; depth, width, spacing and orientation of channels/grooves) used on concrete paved surfaces.
- Megatexture ($\lambda = 50$ to 500 mm, A = 0.1 to 50 mm). Texture with wavelengths in the same order of size as the pavement–tire interface. It is largely defined by the distress, defects, or "waviness" on the pavement surface. Wavelengths longer than the upper limit (500 mm) of megatexture are defined as roughness or unevenness [42]. Figure 4.3 illustrates the three texture ranges, as well as a fourth level—roughness/unevenness—representing wavelengths longer than the upper limit (500 mm) of megatexture.

According to the NCHRP report, there are four main factors affecting the friction between tire and pavement surface [40, 79]. Those four factors are pavement surface characteristics, vehicle operating parameters, tire properties and the environment. In particular, pavement surface characterized by microtexture, macrotexture, megatexture, material properties, and temperature. Skid resistance plays a significant role in wet-skid accidents [28, 40, 47, 79]. Since then, many countries came up with a skid resistance standard to increase safety. Singapore applied a high level of skid resistance to ensure safety within their roadways.

4.1.1 Standard Requirements for Skid Resistance

Roadway authorities are implementing standards for their new roads to guarantee the safety of the motorist. The standard related to skid resistance is mainly based on friction which is derived from the texture (microtexture and macrotexture). Refer to the Code of Practice (COP) for Works on Public Streets [49] and Materials and Workmanship specification [50], Singapore applied a high-level standard of skid resistance for their roadways. The skid resistance for rigid pavement and asphaltic pavement the minimum limitation is 55. The intervention phase will be taken when the skid resistance value reaches 55 for the expressway and 45 for both major and minor roads. All values are based on the British Pendulum Number (BPN). According to the COP, concrete surface paving on footpath shall be approved sprayed-on acrylic modified UV stable pigmented cementitious topping with a minimum thickness of 1 mm, and skid resistance not less than 50. Concrete surface paving on carriageway shall be approved spray-on acrylic modified UV stable pigmented cementitious topping with a minimum thickness of 1 mm, and skid resistance not less than 65. Yet, different authorities apply different levels. Those various standards are summarized in Appendix D.

4.1.2 Test Methods on Skid Resistance

4.1.2.1 Friction Based Tests

There are two types of skid measurement equipment based on their mechanism, namely high-speed friction measurement equipment and low-speed or stationery, that require lane closure. High-speed friction measurement equipment is commonly categorized into four types, i.e., locked wheel testers, side force devices, fixed slip devices and variable slip devices, while three alternatives can be found for the low-speed measurement methods, which are stopping distance measurement, deceleration rate measurement and portable testers. Details can be seen in Appendix E.

Locked-wheel testers can be run with a typical speed of 64 km per hour, and they measure the resistive drag force, and the wheel load with the braking system is fully locked. In order to get the friction number (FN) or skid number (SN), this test can be equipped with either a ribbed tire [13], which is more sensitive to the microtexture surface or a smooth tire [14] which is better to measure the macrotexture. Even though this test is relatively simple, it can only be used on straight segments.

The side force method measures the cornering force perpendicular to the direction of travel of one or two skewed to compute the Mu Number or the sideways force coefficient (SFC). Similar to the locked-wheel tester, it can be run at the same speed. The measurement is continuous, which gives a good result. However, this test is susceptible to road irregularities that can destroy tires quickly. This method is associated with ASTM E670 [15].

The fixed-slip method measures the resistive drag force and the wheel load applied to the pavement, which is used to compute the coefficient of friction, μ or FN as a result of rotational resistance of smooth tires slipping at a constant slip speed (12–20%). The advantage of this method is a continuous measurement and offer high-resolution friction data, but this method also has its disadvantage, such as it uses large amounts of water and requires skilful data reduction.

Variable-slip devices measure skid resistance as a function of slip (range between 0 and 100%) [6]. The result of this method can be related to longitudinal slip friction number, peak slip friction number, slip ratio, estimated friction number and Rado shape factor. However, this method required complex equipment with high maintenance costs and complex data processing.

Both stopping distance measurement (SDM), which is associated with ASTM E445 [12], and deceleration rate measurement (DRM), which is associated with ASTM E2101 [8], require traffic lane closure. During SDM measurement, the wheels of the vehicle must be locked, and the distance the vehicle travels while reaching a full stop is measured. While in DRM measurement, the brakes are applied to lock until deceleration rates can be measured.

The Dynamic Friction Tester (DFT) has the added advantage of measuring the speed dependency of the pavement friction by measuring friction at various speeds. The DFT has three rubber sliders that are spring mounted on a disk at a diameter of 350 mm. The disk is initially suspended above the pavement surface, and it is driven by a motor with a speed of 90 km/h. Water is then applied to the test surface. The friction force is measured by a transducer as the disk spins down. The Dynamic Friction Tester (DFT), which is in accordance with ASTM E1911 [7], is highly portable and easy to handle.

4.1.2.2 Texture Based Test

Two tests based on texture commonly used to measure the skid resistance are the British Pendulum Tester (BPT) and volumetric based spot test, which is called the Sand Patch Method (SPM) test.

BPT, which is in accordance with ASTM E303 [11], is firstly used in the 1960s. The device measures the frictional properties by determining the loss in kinetic energy of a sliding pendulum or rotating disc when in contact with the pavement surface. The loss of kinetic energy is converted to a frictional force and, thus, pavement friction. The typical slip speed for the BPT is usually assumed about 10 km/h. The test result is reported as the British Pendulum Number (BPN). This method provides a measure of microtexture and can be used either in the field or the laboratory. A minimum of five tests should be performed for each specimen.

The SPM test is a volumetric based spot test method that is based on ASTM E965-96 [16] can be used to determine the (average) Mean Texture Depth (MTD) of pavement. This method is widely used as it is practical and inexpensive. The test procedure involves spreading a known volume of material (sand or glass bead) on a clean and dry pavement surface, measuring the area covered, and then calculating the average depth between the bottom of the pavement surface voids and the tops of surface aggregates. The method is significantly influenced by pavement macrotexture. Pavement aggregate particle shape, size, and distribution are not addressed in this method.

The following equation then calculates the average pavement macrotexture depth:

$$\mathrm{MTD} = \frac{4\mathrm{V}}{\pi D^2} \tag{4.5}$$

where

MTD mean texture depth of pavement macrotexture, mm
V sample volume, mm^3
D average diameter of the area covered by the material, mm.

Other tests to measure skid resistance based on texture are Outflow meter, Circular Texture Meter, Texture Depth Gauge, and Electro-optic method test. All of these tests are summarized in Appendix E.

The Land Transport Authority of Singapore (LTA) regularly conducts road condition surveys to monitor the riding quality and skid resistance of major roads and expressways to provide a safe and smooth driving experience. LTA maintain their road network at an acceptable level of service. Three main parameter, including the structural, the functional, and the safety are checked. LTA use a deflectograph to assess the residual life for the pavement by checking the road deflection. The deflectograph is operated at a low speed of 2.5 km per hour.

LTA uses a vehicle with a multifunction road monitor to check the road riding quality. The vehicle consists of a device to measure the pavement roughness (IRI) and the laser profilometer to determine the undulation of road surfaces. In addition, LTA uses a Sideway-force Coefficient Routine Investigation Machine (SCRIM) to measure the wet skidding resistance level of road surfaces.

4.1.3 Treatment Methods to Improve Skid Resistance

Various methods to improve the skid resistance of pavement can be done. It depends on the type of pavement, either asphalt or concrete surface pavement. Examples of friction restoration are chip seal, micro-surfacing, HMA overlay, (longitudinal or transversal) tining, Next Generation Concrete Surface [38], and (longitudinal or transversal) grooving. However, it should be noted that any improvement method will be mainly determined by microtexture and macrotexture. Details can be seen in Appendix E.

A study reported different impacts on the coefficient of friction and sliding velocity caused by the combination of microtexture and macrotexture, whether its low texture and high texture [31]. Furthermore, the size, angularity, shape and texture of aggregate have a different impact on the pavement friction. According to NCHRP Report [40], the angularity and the shape of fine aggregate defines the surface microtexture. The angularity and the shape of coarse aggregates affects its macrotexture, while the texture defines its microtexture which makes some impacts on its friction (Table 4.2).

Table 4.2 Effect on pavement friction

Aggregate fraction	Aggregate property	Effect on pavement friction	
		Asphalt surface	Concrete surface
Fine	Angularity and shape	No effect (–)	Defines pavement microtexture, which highly impacts friction
	Texture	–	Little to no effect
Coarse	Angularity and shape	Defines pavement macrotexture, which significantly impacts friction via hydroplaning potential	If exposed, helps define pavement macrotexture, which significantly impacts friction
	Texture	Defines pavement microtexture, which significantly impacts friction	If exposed, helps define pavement microtexture, which significantly impacts friction

There are two aggregate types commonly used in Singapore for road construction, i.e. granite and steel slag aggregate [33]. Granite is used for asphalt roads since this material is the most common type of aggregates that can be found in the Southeast Asian region. However, the long-term skid resistance performance of granite is only marginally satisfactory. Therefore, since the early 1990s, steel slag has been introduced as an alternative material for road paving purpose. The test results revealed interesting insights into the skid resistance behaviors of the two aggregate types. It was found that water wetting, aggregate gap width, and wheel polishing action all had significant impacts on the skid resistance of the two aggregate types. The skid resistance behaviors of the two aggregates, however, were different under the influence of these factors. The skid resistance of the granite aggregate was affected more by wetting, while the steel slag aggregate deteriorated more as the aggregate gap width widened. The effect of wheel polishing was affected by the aggregate gap width. Results show that steel slag produces a better skid resistance in comparison to granite, with a maximum BPN of 46 and 35 (in wet condition) for steel slag and granite, respectively.

A study was conducted by Lee et al. [53] to check the effectiveness of some common surface texturing methods adopted in Singapore. The methods were studied include troweling and brushing: soft and hard, both longitudinal and transversal direction. They found that troweling, hard, and soft brushing showed a significant improvement in providing skid resistance in dry and wet condition. However, hard and soft brushing methods provide a higher skid resistance than the trowelled surface, especially under wet condition.

As stated from the LTA's website, Kranji Expressway (KJE) was one of the first roads in Singapore (constructed in 1990 and was completed in 1994) that has been introduced to the anti-skid surfacing. It has a total length of 8.4 km from the Bukit Timah Expressway (BKE) to the Pan-Island Expressway (PIE) [51]. The product that has been used to improve the skid resistance in KJE was the use of a special tar-like asphalt substance as a surface material. Further, LTA mentioned that steel slag is their first choice for some road construction projects due to its technical properties, and also because the lifetime road maintenance and replacement costs are minimized. LTA has been using steel slag aggregates since 1990. Using this aggregate provides significant advantages such as high skid resistance, better durability, and resistance to rutting. Having a durable road will also cut carbon emissions due to less road replacement and maintenance [52].

There are two different levels of improving skid resistance. The first level is texturing for new construction, while the second is re-texturing for maintenance or rehabilitation to improve the skid resistance. The table in Appendix E summarizes the texturing techniques. It is adopted and modified from NCHRP report number 108 [40].

4.1.4 Numerical Models on Skid Resistance

The numerical model can be used as an additional tool to analyse the skid resistance. Time cost and resources can be significantly saved by developing a calibrated numerical model. Lee [54] explains many studies related to pavement surface texture were done before, with more studies related to tire design and modelling compared to studies on skid resistance.

Moore et al. [63] pioneers a simplified 3D elastic FEM in relation to rubber slider to study the skid resistance of grooved road pavements. The author concluded that in low sliding speeds, the comparison between experimental measurements of dry sliding friction and elastic theoretical predictions over the grooved pavement could be predicted. A 3D FEM simulates the BPT using ABAQUS. Three different pavement surface types were modelled, both for dry and wet condition. The kinetic energy loss during the test was calculated for comparison with experimental measurements. This model can be used to determine the skid resistance value and contact stresses between the rubber slider and pavement surface. However, the model cannot be used to analyze complex surface textures with non-symmetric patterns [57]. Lee [54] developed A 3D model using ABAQUS. The hyperelastic behavior of the rubber slider was modelled based on experimental test data. Three key issues associated with the BPT were addressed: effects of test surface curvature, effects of slider wear, and unreliable test behavior on coarse-textured surfaces. FEM was used to study the effect of pavement surface texture on BPT [55].

4.2 Surface Water Drainage Performance

The pavement drainage factor is a crucial consideration in cross-section design because water depth has a critical influence on the skid resistance and tire

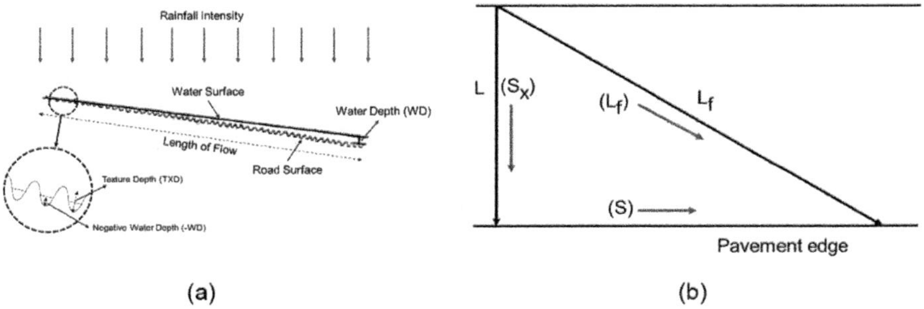

Fig. 4.4 Water flow path **a** over a pavement surface **b** the diagram [2]

hydroplaning resulting from the excess water on the surface pavement. Various parameters: average texture depth, pavement width, pavement super-elevation and grade, and rainfall intensity are affecting water depth above the pavement.

Rainfall excess which is not infiltrated into the pavement will comprise direct runoff to the shoulder lane and side drain eventually. It is common that any rainfall excess should be removed from the surface pavement quickly as the existence of any water above the surface might reduce the grip performance between the tire and the pavement. Two methods are used to define the response of a watershed to a rainfall event, a method based on time of concentration and a quantity-based method (peak-flow and continuous flow models) [65].

Another factor is the flow of the water on pavements. When the rain falls on a sloped pavement surface, it forms a thin film that will increase in thickness as it flows to the edge of the pavement. As can be seen from Fig. 4.4, the thickness of the water surface (water depth) relies on texture depth and the length of the flow path. As the length of the flow path increases, the water depth increases and eventually will cover the pavement macro-asperities. The flow can be considered as channel flow.

The path that the runoff takes to the pavement edge is called the resultant flow path. The correlation between the resultant flow path, length of the flow path and the resultant slope can be determined from the following formula.

$$S_r = (S_x^2 + S^2)^2 \tag{4.6}$$

and

$$L_f = L\left(\frac{S_r}{S_x}\right) = L\left(1 + \left(\frac{S}{S_x}\right)^2\right)^{0.5} \tag{4.7}$$

where

S_x cross slope, m/m
S longitudinal gradient, m/m

S_r slope of resultant flow path, m/m
L_f length of flow path, m
L pavement width, m.

Based on the diagram illustrated in Fig. 4.4a, the depth of water can be computed. Rainfall intensity, length of the flow path, surface texture and the resultant surface slope are the inputs for the formula. As for the resultant surface slope, the parameter of the longitudinal grade significantly influences the flow path length. However, it does not appreciably affect the water depth. The formula can be seen below.

$$WD = 0.01485\left(TXD0.11^{L_f^{0.43}} I0.59^{S_r^{-0.42}}\right) - TXD \quad (4.8)$$

where

WD the water depth above the top of the surface asperities, mm
TF length of flow path, m
TXD texture depth, mm
I average rainfall intensity (mm/hr)
S_r slope of resultant flow path, m/m.

It can be seen from Fig. 4.4b, for a given cross slope (S_X), as the longitudinal grade (S) is steepened or as the pavement width (L) is increased, the flow path (L_f) is lengthened. The resultant flow path depends on the pavement width, cross slope and longitudinal gradient. Guven and Melville [39] did a review on pavement cross slope design related to the surface drainage using climate conditions and design practice in Alabama. Based on his review, he concluded that the primary geometric factors which influence the water depth are the width of the pavement and the cross slope, while the longitudinal gradient has a small influence.

Gallaway et al. study the effects of rainfall intensity, pavement cross-slope and surface texture in relation to the drainage length on pavement water depths [35]. The experimental study found that increasing the surface texture resulted in a decrease in water depth. Greater water depths can be caused by longer drainage lengths and higher rainfall intensities. While increasing pavement cross slope can reduce the water depths. Another conclusion is an increase in surface texture, and pavement cross slope can decrease the water depth detention. A similar finding also mentioned [60] that the increase of the cross slope of the drainage could increase the drainage ability of the pavement.

The Highway drainage guidelines [2] stated that good surface drainage of highway pavements is essential for traffic safety and reasonable maintenance of the roadway surface is necessary. Any water above the surface pavement might interrupt the traffic, reduce the grip, increase the risk of hydroplaning and limit the visibility due to splash and spray. Regarding traffic safety, the grip between the tire and the pavement needs to be optimally fulfilled to minimize the risk of skidding accident.

Yet, there is no research on surface water drainage performance on ECC pavement. As the ECC has a different matrix compared to the conventional concrete, its surface texture and grooves will affect the surface water drainage performance differently. Therefore, research is needed to study the performance of the surface pavement made of ECC.

Further, Singapore's high rainfall intensities bring additional reason to design the surface pavement more properly. Singapore experiences about 2400 mm of precipitation annually with up to 19 rainy days within a month with high rainfall intensity [70]. By 2012, the average of hourly rainfall intensities is about 120 mm which might increase to 140 mm by 2050. Therefore, the post-development runoff has to be engineered efficiently to deal with the high rainfall intensity [43].

4.2.1 Standard Requirements of Surface Water Drainage

Singapore is currently highly developed with many high-rise buildings, rail tracks, roadways, and parking lots, bringing consequences of impervious surfaces. As a result, during the rainy time, peak runoff from the urbanized (post-development) catchment may cause a flash flood. According to the Ministry of National Development of Singapore, Singapore plans to use more than half of its land (58%) for housing, services, jobs facilities, amenities, as well as green spaces and utilities by 2030. While 19% of their land will be used to support mobility and another 19% for defense needs [74]. Since infiltration may have limited performance, especially when pavement will be made of ECC material, then the runoff management should be well engineered. In accordance with the Drainage Handbook [71], the Public Utilities Board (PUB) of Singapore, has established a Code of Practice (COP) on Surface Water Drainage that regulates the minimum engineering requirements for the planning, design and construction of the drainage systems in Singapore. PUB has also developed the guidelines for the implementation of environmentally sustainable green features (including a strategic stormwater management strategy) for users, called the Active, Beautiful and Clean (ABC) Waters. PUB uses the Rational Formula to compute the peak runoff. The runoff coefficient depends on the degree and the type of development within the catchment. For roads, highways, airport runways and paved up areas should use a value of 1.00 for C.

$$Q_r = \frac{1}{360} CIA \tag{4.9}$$

where

Q_r peak runoff at the point of design (m^3/s)
C runoff coefficient
I average rainfall intensity (mm/hour)
A catchment area (hectares).

PUB specifies a time range from 5 to 15 min for the overland flow time (to), depending on the overland travel distance, land topography and characteristics to

Table 4.3 Grade design criteria by LTA Singapore

Criteria	Express-way	Semi-express way	Arterial road		Other road	
			Major	Minor	Primary access	Local access
Minimum grade (%)	0.4[a]	0.4[a]	0.4[b]	0.4[b]	0.4[b]	0.4[b]
Desirable max grade (%)	4	5	5	6	6	6
Absolute max grade (%)	5	6	6	8	8	8

Notes
[a] A road gradient of less than 0.4% may be used for widening of expressway
[b] For built up area, a road gradient of less than 0.4% may be used to tie in with existing access

calculate the time of concentration. The drain flow time (td) can be estimated from the hydraulic properties of the drainage channel.

Ideally, a runoff will be highly dependent on rainfall intensities, cross fall, longitudinal gradient, road surface texture and pavement width. To facilitate this issue, LTA (2010) specify a requirement for a cross slope or cross-fall at grade road, 1 in 36 and 1 in 30 for absolute minimum cross-fall and desirable cross-fall, respectively. The requirement for the grade of a carriageway can be seen in Table 4.3.

In other countries, they set a regulation regarding the cross-fall and grade, and other parameters, such as flow path length. The flow path length is the distance of water flowing on the road, an important factor for road surface drainage. As an example, Japan requires a flow path length that is not greater than 40 m.

Regarding water (surface drainage) outflow, various standards in the USA specify a requirement for lower outflow times. The Texas DOT [76] and the State of North Carolina DOT [75] set a regulation of an average of 10 s or less shall be obtained for any lane mile section, while the Louisiana DOT [59] specifies that an average outflow reading of greater than 9.00 s. This study investigates various methods to improve surface water drainage. Tests with an outflow meter were conducted.

Textured surfaces not only reduced the splash and spray but also help to diffuse the light to decrease the headlight glare from an oncoming vehicle. The headlight glare is increased at night during wet condition. Standard ASTM 793 [5] recommended a minimum texture depth between 1 and 1.5 mm accordingly.

4.2.2 Test Methods on Surface Water Drainage

According to ASTM E2380M-15 [9], an outflow meter test can be conducted to evaluate the drainage performance of the pavement through its surface and subsurface voids. It measures the duration of a known quantity of water taken to run through voids in the pavement texture and intended to provide a measure of the ability of the pavement to relieve pressure from the surface of the tires, thus indicating hydroplaning potential under wet conditions. A faster run time indicates a thinner film of water may exist between the tires. Minimum five tests need to be performed for each sample.

4.2.3 Treatment Methods to Improve Surface Water Drainage

Surface water drainage is significantly affected by cross slope and texture depth [34]. The direction of texturing less significantly influences it. Thus, it is suggested to construct sufficient deep textures. Then secondly, it comes the direction as the NCHRP study concluded that grooving could be an alternative for better drainage when the direction of the groove is in line with the gradient [42].

A proper design of sag-vertical curves can reduce hydroplaning. This design will support by maximum pavement width of 11 m, maximum inlet spacing (15–61 m), channelized curb flow encroachment, and increase in roughness coefficient by 0.005 with Manning's [34]. The splash and spray, which are closely related to water on surface drainage, are affected by macrotexture. Thus, by improving the macrotexture of the surface pavement, the surface water drainage will be improved. In details, the angularity and shape of coarse aggregate will help to improve the surface water drainage.

Several texture methods can be done to improve surface water drainage. Transverse grooving for concrete pavement is known as the most direct drainage channel of water of the pavement even though it is not common due to construction difficulties. Alternatively, longitudinal grooving also does provide a channel for water and produces a sufficient tracking effect on tires (vehicles), thus reducing the risk of skidding accidents [24]. Porous asphalt/concrete pavements are alternative for pavement application regarding water storm [61]. Porous pavements mixtures consist of binder, additives, and a high proportion of coarse aggregate with limited fines, resulting in a high percentage of permeable air voids (typically 15–20%), thus allowing water to penetrate through the material. The high air void content creates permeability, which decreases the water on the roadway, help to reduce the splashing and spraying.

4.2.4 Numerical Models on Surface Water Drainage

Since there is no previous surface water drainage study on ECC, the numerical model can be utilized to study the rainfall loading on a highway in Singapore. Various textures, cross falls, and slopes will affect the surface drainage performance of the new pavement differently. The numerical model can be done by using fluid dynamics. However, until now, there are not many numerical models on water surface runoff. Many numerical models were developed on permeability and hydroplaning. Various investigations related to these three subjects are summarized in Tables 4.4, 4.5, and 4.6.

Table 4.4 Numerical model on water surface runoff

Numerical model and tool	Remarks
Pavement drainage Using Geoslope [60]	The FEM model were designed to evaluate the effect of (1) broken and seated concrete layer with or without a Superpave asphalt layer, (2) central collection pipe, (3) Superpave HMA surface, (4) pavement geometry, (5) cracks on pavement surface and (6) slope of drainage blanket
Pavement surface runoff model using DUNE and DUMU, which were developed by University of Stuttgart [80]	Flow resistance in the Pavement Surface Runoff Model is described by the Darcy-Weisbach equation based on the depth-averaged shallow water equations. The model includes a model road section with drainage facilities, an existing road section with irregular topography as well as a road section designed in CAD with a complex boundary

Table 4.5 Numerical model on permeability

Numerical model and tool	Remarks
2D FEM analysis using GEOSLOPE [60]	The analysis of permeability on Superpave surfaces had a moderate effect on the pavement drainage. While permeability of AC base had a significant effect on the pavement drainage
3D FEM model of pavement subsurface drainage systems using FlexPDE program [58]	By conducting the unsaturated transient analyses, the drainage characteristics corresponding to various drainage design practices, geometry parameters and material properties were identified
2D FEM model on pavement infiltration using HYDRUS-2D [44]	The simulation results of the volumetric water content θ in a pavement construction HYDRUS-2D has been applied on a pervious pavement with slots of 10 mm on a 40 cm base layer of gravel (0/45 mm) for a rain intensity of 100 l/(s·ha)
3D FEM model of pervious concrete pavement using ABAQUS [3]	The porosity distribution model was validated by comparing the results from EverFE. Three layers pervious concrete model developed with assumptions of linearly elastic material, the modulus of elasticity of pervious concrete is a function of the porosity, The Poisson's ratio is constant along the depth of the pervious concrete cross section, the porosity distribution is constant in the longitudinal and transverse directions of the pavement

Table 4.6 Numerical model on hydroplaning

Numerical model and tool	Remarks
A rubber sliding on a two-dimensional smooth sinusoidal asperity separated by a thin fluid film is modeled [62]	The first analytical treatment of the hydroplaning problem was proposed by [62]. A one-dimensional Reynolds equation solution was obtained in which inlets, central and outlet regions for the fluid film were treated separately
2D model on a tire deformation for hydroplaning [23]	Using the Navier–Stokes equations, inviscid, laminar, turbulent models and side flow were explored. However, the recovery factor is lower compare to NASA's experiment
3D FEM model of the dynamic aquaplane problem using ADINA [83]	Simulation of the fluid flow with the consideration of deformable tire structure
Simulation of hydroplaning through fluid–structure interaction using MSC.DYTRAN [66]	The simulation has a drawback in that the fluid flow was modeled using the potential flow theory. Pavement surface characteristics was not properly developed and its influence on hydroplaning was not analyzed
3D simulation of effects of viscous hydroplaning using MATLAB for meshing and ABAQUS [4]	Two different elasticity models of car tyre were modeled to simulate the viscous hydroplaning effects. Smooth surfaces were assumed, i.e. zero microtexture depth
Simulation of hydroplaning based on 3D model using FLUENT (Finite volume method) [67]	Based on Navier–Stokes equation, the 3D model developed using pneumatic tire model, fluid flow model and the pavement surface model with consideration of tire-pavement contact and tire-fluid interaction
Study of 3D model of hydroplaning risk on rolling and sliding car using ABAQUS [48]	Smooth tire was considered to roll and slide over plane pavement surface. Hydroplaning speed calculated using the developed simulation model was duly validated against the past experimental studies and the results show a good agreement with the experimental findings for both rolling and sliding cases and with the well-known NASA hydroplaning equation

The numerical model with liquid dynamics tool can be used to study how material and textures can affect the surface water drainage performance. With numerical tools, a model with various textures can be created. Therefore, finding the optimum solution for safe pavement can be done without doing many experimental tests, which need a lot of resources and costly.

Table 4.7 Noise limit in Singapore

Affected premise	Max noise limit (dB)		
	Day (07:00–19:00)	Evening (19:00–23:00)	Night (23:00–07:00)
Residential	65	60	55

4.3 Tire-Pavement Noise

Singapore has transformed into a highly developed city with most of the population live in urban areas. Due to limited space, the most residential area is developed close to expressways and major arterial roads with high traffic densities which cause noise pollution. Strait times [56] reported that communities have public complaints about noise pollution. According to Gibbs et al. [37], tire-pavement noise represents 75–90% of the total noise generated by passenger vehicles and trucks. Tire-pavement noise is the main factor affecting the overall traffic that should be reduced to have quieter pavement systems. Three parameters of pavement properties are significantly affecting tire-pavement noise, namely texture, porosity, and stiffness [72].

4.3.1 Standard Requirements for Noise

The NEA of Singapore implemented residential noise limits for different periods of the day, which described below. The noise limit of 67 dB is applied to new roads and expressways (Table 4.7).

4.3.2 Test Methods on (Tire-) Pavement Noise

Commonly, there are two ways to measure noise: wayside noise and source noise measurements. Wayside noise measurement is done at a fixed distance from the road (at the side of the road). The measurement is using microphones which is ideally placed between 7.5 and 15 m away from the road. Source noise measurements intended to study more about tire-pavement noise.

There are three common types of wayside testing. In instances where there is little traffic, statistical pass-by (SPB) testing can be done. In this case, a microphone at a fixed position is used to measure the maximum sound levels (Lmax) of hundreds of individual vehicles. Then, a calculation is made of the sound level from an "average" car, medium truck, and heavy truck travelling at a standardized speed. Second, a test of vehicle/tire combinations called a controlled pass-by (CPB) test. Using SPB and CPB, pavements at different locations can be compared to one another. Some caution must be exercised, however, as there can always be differences in the "average" vehicle from site to site (for SPB) or unique interactions between a specific tire and pavement combination (for CPB). The third type of wayside testing that is commonly conducted is termed time-averaged, sometimes referred to as continuous flow traffic time-integrated model (CTIM). In this case, the microphone is set to record all of the traffic noise over a fixed time (commonly 5–30 min), and traffic levels and speeds are simultaneously recorded. An average equivalent sound level (Leq) over this period is calculated and is often reported as an average of repeat measurements. There are two types of noise measurements on driving vehicles under well-controlled conditions, including (1) the Pass-by situation where is characterized by the fact that the engine is on during the passing of the vehicle, and (2) the Coast-by situation indicates the engine-off [64].

There are two principal techniques for measuring tire-pavement noise: close-proximity (CPX) and on-board sound intensity (OBSI). General Motors initially developed OBSI for tire evaluation at their test facilities. The technology was developed further by Donavan [29] of Illingworth & Rodkin as part of quiet pavement research for the Departments of Transportation in both California and Arizona. The CPX method uses single microphones that measure sound pressure. OBSI uses dual-microphone probes that measure sound intensity. The latter is directive, meaning that the measurements from each of the two microphones can be used to sort out the direction of the various sources. Currently, the CPX and OBSI methods specify different positions for the microphone, including the height (from the ground), the spacing between the front and rear positions, and distance from the tire sidewall.

The Tire-Pavement Test Apparatus (TPTA), which pioneered at Purdue University, consists of a rotating arm with a tire assembly that is in contact with individual test panels that allow real pavement materials to be tested for tire-pavement noise at up to 30 mph. Microphones and probes can be positioned to collect CPX or OBSI style measurements. While large in size and used primarily for screening potential quieter pavement types in a research environment, the TPTA serves as a model for potential lab tests of the future to measure tire-pavement noise more directly prior to full-scale implementation. Noise absorption measurement can be considered to study the effect of texture and porosity, as this technology is reliable, handy, and it can be used in the laboratory and in situ [78].

A noise investigation was conducted in an apartment block adjacent to the East Coast expressway. The apartment block is 50 m away from the expressway. It is a major expressway (10-lane) linking the international airport and the eastern areas of the island with the central business district, port and the west coast of Singapore. Noise data was logged using a noise data sensor and data logging software. The noise sensor was mounted near an open window facing the expressway on the 11th floor apartment. Noise data was logged from 08:30 a.m. to 10:00 p.m. every night for all days of the week. The logging software was set to sample noise data every minute. It was found out that most of the highways produced higher noise more than the limits specified by the Singapore government [17].

4.3.3 Treatment Methods to Reduce Noise on (Tire-) Pavement

According to Little Book Quieter Pavements, three main parameters are considered for quieter pavements: texture, porosity, and stiffness. Firstly, texture has to be remaining small (less than 5 mm) and negative. Secondly, it is better to have higher porosity as it does help to absorb noise and reduce the contact area. It works better when the porosity is more than 20%. However, air voids might affect durability. Thus, the porosity has to be carefully designed. Lastly, the stiffer the pavement, the quieter it is [72].

Porous asphalt and gap-graded stone matrix asphalt (SMA) are typically used for quieter asphalt pavements. In Europe, double-layer porous with maximum aggregate size 6–8 mm is one common type used for quieter pavement. In Arizona (USA), the asphalt rubber friction course (ARFC) used in many locations. It is an open-graded material with an additional binder. Research done by the National Concrete Pavement Technology Center has identified the quieter concrete pavement options that do not compromise on safety factor. Over one thousand pavement test sections throughout the US and Canada have been tested. As results, drag surfaces (burlap and artificial turf) and diamond ground surfaces are among the quietest. Then, longitudinal tining also offers quieter pavement. In Europe, a common method used for concrete pavement surfacing is the exposed aggregate and porous concrete [45, 46].

4.4 Texture Durability

Texture durability is one of the critical points for the pavement to evaluate the effects of aggregate quality on microtexture and macrotexture loss overtime/traffic. The study gives an understanding of how concrete mixtures with tougher, more durable aggregates retain higher friction values loss over time/traffic. From the same study of texture durability, microtexture deterioration was investigated by using DF Tester. As an example, the usage of high-silica granite, which was used in Colorado, while granite was used in Minnesota shown high initial DFT(20), and it remains high after high traffic. The use of limestone in Kansas and Illinois shown a greater rate of microtexture deterioration. It was concluded that the use of higher quality aggregates in the concrete mixture helps to maintain the microtexture qualities needed for friction [40]. In conclusion, the most significant loss of macrotexture over traffic is experienced by diamond-ground textures. Meanwhile, the lowest loss is for longitudinally grooved and dragged textures. The geometric shape of the diamond ground textures results in more substantial loss than textures with no grooves.

4.5 Self-cleaning Function on Pavements

The projected Business-as-usual (BAU) emissions in Singapore are to reach 77.2 million tons in 2020. Thus, Singapore is aiming to reduce its carbon emissions up to 11% below business-as-usual-scenario levels by 2020. The primary emissions from traffic are nitrogen oxides (NO_x), hydrocarbons (HC), and carbon monoxide (CO), causing a significant impact on the urban quality such as urban smog, acid rain, fine particle pollution, and tropospheric ozone, which causing adverse issues for human health and the environment [18]. One way to fight the emission is through the innovative (material for) pavement application. The self-cleaning function on pavements can be enabled with the application of photocatalytic materials. Titanium dioxide (TiO_2) is used to provide an air-purifying surface to reduce air pollution from traffic. When it is used in urban areas, where the risk of smog formation is increased during hot sunny days, the air pollution will be decreased significantly. The commercialization of TiO_2-based photocatalytic products were started in the mid-1990s in Japan. Nowadays, the industry has grown rapidly as more than 2000 companies have emerged since 2003. Fujishima and Zhang [32] categorized the related industry into five: exterior construction materials, interior furnishing material, road construction materials, purification facilities, and household goods. Among those five categories, self-cleaning exterior construction materials occupied the biggest market share.

4.5.1 Photocatalytic Process

The photocatalytic process occurs in a way a semiconductor activated by light at a particular wavelength creates hydroxyl radicals (*OH) and superoxides (O_2^-), which accelerate the natural decomposition process of organic pollutants. As results, the pollutants are decomposed as non-hazardous waste products. Afterwards, the non-hazardous pollutants will be removed from the surface pavements by the washing rain (Fig. 4.5).

Heterogeneous photocatalysis had shown promising results for NO_x abatement. Ultraviolet (UV)-light was commonly used to activate the photocatalyst. Recently, a new source of light, visible light (VL), is becoming a new way. Visible light composes the largest past of solar radiation, which can be an advantage as the content of UV light in indoors is significantly smaller than that in sunlight environment because the fluorescent lamp mainly emits visible light. This means that photocatalytic application can also be applied in tunnels and indoor environments. It is worth noting that the air pollutants in tunnels are normally in high concentration.

Three methods can be used on the utilization of visible light for photocatalysis: (1) utilization of semiconductors to absorb visible light, such as CdS, Cu_2O, Fe_2O_3, and WO_3, (2) doping of photocatalysts with metal ions (such as Cr^{3+}, Fe^{3+}) to create local energy levels within the band gaps of visible spectral range, and (3) by doping of photocatalysts with non-meal atoms, such as N, S, and C. The third method claimed

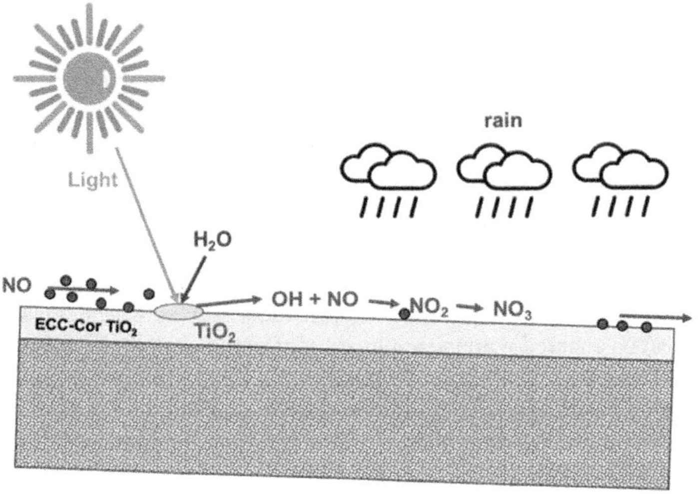

Fig. 4.5 Photocatalytic process on concrete

to be the most efficient method for the developing photocatalysts utilizing visible light [32].

4.5.2 Test Methods on Photocatalytic Performance

Test methods on photocatalytic performances are varied depending on the type and size of TiO_2, the source lights and pollutant solutions. One example is the examination via decomposition of Rhodamine B (RhB), which represents the pollutant solutions associated with dirt and stains, is one of the most common dyes used (UNI 11259:2008). Then the efficiency of photodecomposition of the sample can be observed with a spectrophotometer by measuring the light reflectance at a different wavelength. The reflectance of the sample is tested before and after it is exposed to dye and the lights. Cassar et al. [25] have investigated white cement containing TiO_2 Degussa P25 with yellow dye (Phenantroquinone) under 400 W simulator lamps with $\lambda > 290$ nm. Vallee et al. [77] investigated white cement mixed with T_iO_2 anatase with RhB dye under 300 W UV light Ultra-Vitalus. Bolte [19] was tested mortar made with cement mixed with Degussa P25a with RhB dye under 100 W mercury vapour lamp with $\lambda < 320$ nm filtered out to eliminate photolysis. Yuranova et al. [81] have investigated cement-based plates with T_iO_2/S_iO_2 surface layers as coating with TiO_2 rutile with red wine dye under a sun test solar simulator (290–800 nm). Bonafous et al. [20] have studied the effect of mortar with both white and ordinary cement mixed with T_iO_2 PC 105 with RhB dye under Bromocresol green 2.5 $\mu g/cm^2$ 300 W lamp. Zhao et al. [82] investigated White ECC mixed with P25 with RhB dye under 9 W artificial UV and sunlight. Another test is the dirt pick-up resistance

(DPUR) which is used to investigate the amount of dirt that can be removed from the surface sample when the sample is exposed to the lights.

4.5.3 Research and Application of Photocatalytic Pavements

Various studies have incorporated photocatalytic oxidation in pavement materials for NO_x reduction. Laboratory results indicate a good efficiency towards the abatement of NO_x in the air by using these innovative materials for pavement applications. Yet, real applications on a large scale are still showing inconsistency as there are many parameters that cannot be fully controlled, such as humidity, light intensity, and wind speed [22].

Antwerp's photocatalytic concrete paving blocks in 2004 was one of the first projects embedding photocatalytic function for pavements application. From this project, several insights were gained. Good durability of the efficiency towards NOx abatement was observed. The deposition of pollutants on the surface leads to a decrease in efficiency, which can be regained after washing. Measurements in 2010 indicate that even after more than five years of service life, the photocatalytic efficiency of the pavers is still present [22]. In 2007, Poon and Cheung [69] researched by incorporating T_iO_2 into a concrete paving block with different surface mix designs. The maximum aggregate size of the recycled fine aggregate used is 2.36 mm. Important results from the test show (1) the NO removal ability was increased when the porosity of the block was increased (lower density or material with less fine particles), (2) sample with a lower cement to aggregate ratio leads to a higher efficiency rate for NO removal, (3) P-25 T_iO_2 shows a better photocatalytic performance compare to Rutile, but not in cost perspective [69].

In 2011, a photocatalytic function was embedded using cementitious coating materials in the Leopold II tunnel on project PhotoPAQ, which was funded by the European Life+. The lower result was gained from the expected performance. However, there are many interesting conclusions were drawn. The typical low average wind speed (~3 m s^{-1} during daytime), the cold, and low average relative humidity (70–90% r.h.) inside the tunnel resulted in a reduction of the photocatalytic performance. Additionally, the UVA irradiance, which was below the target (it only reached 0.6–1.6 Wm^{-2}, while the target is to produce more than 4 Wm^{-2}) is affected the photocatalytic performance [36].

Also, in 2011, two trial sections (each 30 m long and 3 m wide) of double-layered concrete have been built with photocatalytic materials. The bottom layer concrete was built with more than 50% of the large aggregates replaced by recycling concrete aggregates. At the same time, the top layer built with aggregate with a maximum aggregate size of 6.3 mm. After conducting the test, it was found out that the curing compound has a negative effect on NO_x reduction. Different surface treatment has been performed: exposed aggregates, smooth (formwork side), and sawn surface. The results show that the surface of the exposed aggregates performs equally well as the smooth (formwork surface) but lower than the sawn surface. This caused by the combination of less photoactive cement at the surface and a higher surface porosity [21].

Three methods of application for T_iO_2 to concrete pavement were investigated: (1) a cementitious-based ultrathin coating, (2) a water-based titanium dioxide solution, and (3) sprinkling T_iO_2 to the fresh concrete surface before hardening [41].

References

1. AASHTO (1993) Design of pavement structures. AASHTO, Washington, DC
2. AASHTO (2007) Highway drainage guidelines: Chapter 9: Storm drain systems. AASHTO, Washington, DC
3. Alam A, Haselbach L, Cofer WF (2013) Three-dimensional finite element modeling of pervious concrete pavement: vertical porosity distribution approach. Int J Res Eng Technol 2(12)
4. Andrén P, Jolkin A (2003) Elastohydrodynamic aspects on the tyre-pavement contact at aquaplaning. VTI report 483A, Linkoping
5. ASTM 793 (1981) Frictional interaction of tire and pavement: ASTM Special Technical Publication 793, Ohio, USA
6. ASTM E1859-E1859M (2015) Test method for friction coefficient measurements between tire and pavement using a variable slip technique. ASTM International, West Conshohocken, PA
7. ASTM E1911 (2009) Standard test method for measuring paved surface frictional properties using the dynamic friction tester, West Conshohocken, PA
8. ASTM E2101 (2015) Standard test method for measuring the frictional properties of winter contaminated pavement surfaces using an averaging-type spot measuring deceloremeter. ASTM International, West Conshohocken, PA
9. ASTM E2380M-15 (2015) Test method for measuring pavement texture drainage using an outflow meter. ASTM International, West Conshohocken, PA
10. ASTM E274/E274M-11 (2011) Standard test method for skid resistance of paved surfaces using a full-scale tire. ASTM International, West Conshohocken, PA
11. ASTM E303-93 (1993) ASTM standard test method for measuring surface frictional properties using the British pendulum tester. ASTM International, West Conshohocken, PA
12. ASTM E445, Test method for stopping distance on paved surfaces using a passenger vehicle equipped with full-scale tires. ASTM International, West Conshohocken, PA
13. ASTM E501 (2015) Standard specification for standard rib tire for pavement skid-resistance tests. ASTM International, West Conshohocken, PA
14. ASTM E524 (2015) Specification for smooth tire for pavement skid-resistance tests. ASTM International, West Conshohocken, PA
15. ASTM E670 (2015) Standard test method for testing side force friction on paved surfaces using the mu-meter. ASTM International, West Conshohocken, PA
16. ASTM E965-96 (2006) Standards test method for measuring pavement macrotexture depth using a volumetric technique (reapproved 2006). ASTM International, West Conshohocken, PA
17. Bhanap I (2013) An analysis of roadway noise at residential estates in close proximity to expressways in Singapore. Noise Health 15(64):183–189
18. Birnbaum R, Price D, Batiuk R, McCullough M, Smith R (1997) Nitrogen oxides: impacts on public health and the environment. EPA 452/R-97-002, Washington, DC
19. Bolte G (2005) Photocatalysis in cement-bonded materials. Cement Int 3(92)
20. Bonafous L, Cassar L, Cassat P, Colombet P, Guillot L (2009) Photocatalytic granular mixture for mortar and concrete and its use
21. Boonen E, Beeldens A (2011) Photocatalytic applications in Belgium, purifying the air through a double layered concrete pavement. In: Second international conference on best practices for concrete pavements, Florianopolis, Brazil, November 2nd to 4th

22. Boonen E, Beeldens A (2014) Recent photocatalytic applications for air purification in Belgium. Coatings 4(3):553–573. https://doi.org/10.3390/coatings4030553
23. Browne A (1971) Dynamic hydroplaning of pneumatic tires, Northwestern University
24. Caltrans (2007) Rigid pavement preservation: Chapter 5: Diamond grinding and grooving, 2nd edn. Caltrans, California
25. Cassar L, Pepe C, Pimpinelli N, Tognon G, Guerrini G, Amadelli R (2003) White cement for architecture concrete, possessing the photocatalytic properties
26. Corsello P (1993) Evaluation of surface friction guidelines for Washington State Highways. Olympia, Washington
27. Data Government of Singapore (2017) Causes of road accidents: causes of accidents by severity of injury sustained. https://data.gov.sg/dataset/causes-of-road-accidents-causes-of-accidents-by-severity-of-injury-sustained?view_id=bd999ce1-8629-4783-a3c3-66d2665fc66e&resource_id=d68321b6-c438-425d-b9f4-d5777eee9e77
28. Davies R, Cenek PD, Henderson RJ (2005) The effect of skid resistance and texture on crash risk. In: International conference for surface friction for roads and runways, May 2005, Christchurch, New Zealand
29. Donavan PR (2006) Comparative measurements on tire pavement noise in Europe and the USA. FHWA/CA/MI-2006/09, California
30. FHWA (2015) Evaluation of pavement safety performance. FHWA-HRT-14-065, Washington, DC
31. Flintsch G, Al-Qadi I, Davis R, McGhee K (2002) Effect of HMA properties on pavement surface characteristics. In: Pavement evaluation conference, Roanoke, Virginia
32. Fujishima A, Zhang X (2006) Titanium dioxide photocatalysis: present situation and future approaches. C R Chim 9(5–6):750–760. https://doi.org/10.1016/j.crci.2005.02.055
33. Fwa TF, Choo YS, Liu Y (2003) Effect of aggregate spacing on skid resistance of asphalt pavement. J Transp Eng 129(4):420–426. https://doi.org/10.1061/(ASCE)0733-947
34. Gallaway B, Ivey D, Hayes G, Ledbetter W, Olson R, Woods D, Schiller Jr R (1979) Pavement and geometric design criteria for minimizing hydroplaning. FHWA-RD-79-31
35. Gallaway BM, Schiller RE, Rose JG (1971) Effects of rainfall intensity, pavement cross slope, surface texture, and drainage length on pavement water depths. 138-5, Texas
36. Gallus M, Akylas V, Barmpas F, Beeldens A, Boonen E, Boréave A, Cazaunau M, Chen H, Daële V, Doussin JF, Dupart Y, Gaimoz C, George C, Grosselin B, Herrmann H, Ifang S, Kurtenbach R, Maille M, Mellouki A, Miet K, Mothes F, Moussiopoulos N, Poulain L, Rabe R, Zapf P, Kleffmann J (2015) Photocatalytic de-pollution in the Leopold II tunnel in Brussels: NOx abatement results. Build Environ 84:125–133. https://doi.org/10.1016/j.buildenv.2014.10.032
37. Gibbs D, Iwasaki R, Bernhard R, Bledsoe J, Carlson D, Corbisier C, Fults K, Hearne T, McMullen J, Newcomb D, Roberts J, Rochart J, Scofield L, Swanlund M (2005) Quiet pavement systems in Europe. FHWA-PL-05-011, Virginia
38. Guada IM, Spinner D, Rezaei A, Harvey JT (2012) Evaluation of grind and groove (Next Generation Concrete Surface) pilot projects in California, California
39. Guven O, Melville JG (1999) Pavement cross slope design: a technical review. 36849-5337, Auburn, Alabama
40. Hall J, Smith K, Littleton P (2009) Texturing of concrete pavements: NCHRP Report 634, Washington, DC
41. Hassan MM, Dylla H, Mohammad LN, Rupnow T (2012) Methods for the application of Titanium Dioxide coatings to concrete pavement. Int J Pavement Res Technol 5(1):12–20
42. Henry J (2000) Evaluation of pavement friction characteristics: a synthesis of highway practice, Washington, DC
43. Hua D, Wen LY (2011) Managing stormwater in urbanised Singapore for flood control. Flood and Storm Surge Control Training, Tokyo, Japan
44. Illgen M, Harting K, Schmitt TG, Welker A (2007) Runoff and infiltration characteristics of permeable pavements: review of an intensive monitoring program

45. Skarabis J, Gehlen C, Stöckert U, Alte-Teigeler T (2014) Noise emission of concrete pavement surfaces produced by diamond grinding. Innovative Solutions—Benefiting Society. In: 12th international symposium on concrete roads, Prague. European Concrete Paving Association, Prague
46. Jurrians G, Keulen Wv, Kramer W (2012) New guidelines for exposed aggregate surfaces in the Netherlands. In: 10th international conference on concrete pavements, Québec, Québec
47. Karim M, Chyc-Cies J, Schick D, Dechkoff C (2012) Evaluation of a skid resistant material at high incident intersection locations, New Brunswick
48. Kumar SS, Anupam K, Scarpas T, Kasbergen C (2012) Study of hydroplaning risk on rolling and sliding passenger car. Procedia Soc Behav Sci 53(3):1019–1027. https://doi.org/10.1016/j.sbspro.2012.09.951
49. Land Transport Authority (2008) Code of practice for works on public streets, 6th edn. Land Transport Authority, Singapore
50. Land Transport Authority (2010) Materials and workmanship specification for civil and structural works. Land Transport Authority, Singapore
51. Land Transport Authority (2013) Kranji Expressway (KJE) projects. http://www.lta.gov.sg/content/ltaweb/en/roads-and-motoring/projects/kranji-expressway-kje.html. Accessed Aug 2016
52. Land Transport Authority (2016) Using steel slag aggregate in construction. https://www.lta.gov.sg/content/ltaweb/en/about-lta/lta-and-the-community/lta-and-the-environment/using-steel-slag-aggregate-in-construction.html. Accessed Aug 2016
53. Lee YPK, Fwa TF, Choo YS (2003) Skid resistance evaluation of concrete pavement surfaces. J Eastern Asia Soc Transport Stud 5
54. Lee K (2004) Analyzing laboratory skid resistance test using finite element modeling. Doctoral dissertation, National University of Singapore
55. Lee YPK, Fwa TF, Choo YS (2005) Effect of pavement surface texture on British Pendulum Test. J Eastern Asia Soc Transport Stud 6:1247–1257
56. Lian GC (2010) Noise pollution in Singapore: when buzz becomes noise. http://wildsingaporenews.blogspot.sg/2010/11/noise-pollution-in-singaporewhen-buzz.html#.WBMRYC196Uk
57. Liu Y, Fwa TF, Choo YS (2003) Finite-element modeling of skid resistance test. J Transp Eng 129(3):316–321. https://doi.org/10.1061/(ASCE)0733-947X(2003)
58. Liu Y (2005) Three dimensional finite element modeling of pavement subsurface drainage systems. Doctoral dissertations, University of Kentucky
59. Louisiana Department of Transportation (2010) Special provision state project no 737-99-1102. Lousiana DOT, Louisiana
60. Mahboub KC, Liu Y, Allen DL (2003) Evaluation and analysis of highway pavement drainage. Lexington, Kentucky
61. Mansour TN, Putman BJ (2013) Influence of aggregate gradation on the performance properties of porous asphalt mixtures. J Mater Civ Eng 25(2):281–288. https://doi.org/10.1061/(ASCE)MT.1943-5533.0000602
62. Moore DF (1967) A theory of viscous hydroplaning. Int J Mech Sci 9:797–810
63. Moore I, Purushothaman N, Heaton B (1988) Three-dimensional elastic finite element study of the skid resistance of grooved pavement. Int J Numer Meth Eng 26:437–452
64. Müller G, Moser M (2012) Handbook of engineering acoustics. Springer, Berlin
65. Nicklow, Boulous, Muleta (2006) Comprehensive urban hydrologic modeling handbook for engineers and planners. MWH Soft, Incorporated, United States
66. Okano T, Koishi M (2001) A new computational procedure to predict transient hydroplaning on a tire. Tire Sci Technol 29(1):2–22
67. Ong Ghim Ping Raymond (2006) Hydroplaning and skid resistance analysis using numerical modeling. PhD thesis, National University of Singapore
68. Permanent International Association of Road Congresses (1987) Report of the Committee on Surface Characteristics. 18th World Road Congress, Brussels, Belgium
69. Poon CS, Cheung E (2007) NO removal efficiency of photocatalytic paving blocks prepared with recycled materials. Constr Build Mater 21(8):1746–1753. https://doi.org/10.1016/j.conbuildmat.2006.05.018

70. PUB Singapore (2013) Managing urban runoff (drainage handbook). PUB and IES, Singapore
71. Public Utilities Board of Singapore (2013) Managing urban runoff. PUB and IES, Singapore
72. Rasmussen RO, Bernhard R, Sandberg U, Mun EP (2007) The little book of quieter pavements. FHWA, Austin, Texas
73. Sandburg U, Ejsmont J (2002) Tyre/road noise reference book. Informex, Kisa, Sweden
74. Singapore Business Review (2019) Singapore needs 5,600ha more land supply by 2030. https://sbr.com.sg/economy/news/singapore-needs-5600ha-more-land-supply-support-69m-population-2030
75. State of North Carolina DOT (2012) Contract proposal division 03
76. Texas DOT (2004) Pavement texturing special specification 7331, 6165th edn. Texas DOT, Texas
77. Vallee F, Bonafous L, Guillot L, Pimpinelli N (2004) Innovative self-cleaning and de-polluting facade surfaces. CIB World Building Congress, Toronto, Canada
78. Van Keulen W, Li M, Ceylan H, Molenaar A, Van de Ven M (2016) SIROTOL: statistical model of tyre-road noise for thin layer surfacing
79. Wallman C-G, Astrom H (2001) Friction measurement methods and the correlation between road friction and traffic safety: a literature review. VTI meddelande 911A, Sweden
80. Wolff A (2012) Simulation of pavement surface runoff using the depth-averaged shallow water equations. Doctoral dissertation, University of Stuttgart
81. Yuranova T, Sarria V, Jardim W, Rengifo J, Pulgarin C, Trabesinger G, Kiwi J (2007) Photocatalytic discoloration of organic compounds on outdoor building cement panels modified by photoactive coatings. J Photochem Photobiol 188(2):334–341
82. Zhao A, Yang J, Yang E-H (2015) Self-cleaning engineered cementitious composites. Cement Concr Compos 64:74–83. https://doi.org/10.1016/j.cemconcomp.2015.09.007
83. Zmindak M, Grajciar I (1997) Simulation of the aquaplane problem. Comput Struct 64:1155–1164

Chapter 5
Multi Criteria Decision Analysis on the Functional Performance of Pavement

This chapter summarizes the results of Multi Criteria Design Analysis (MCDA) on the functional performance of pavement obtained in this thesis. The MCDA proposes a sophisticated method to define the functional performance of surface properties of such a type of surface pavement. The functional performance includes microtexture, macrotexture, pavement drainage (splash and spray), pavement unevenness, noise, cost, and durability.

In most cases, a design requires coupled optimization of two or more parameters of performance. An example given by Ashby [1], a material was needed for a beam for a bicycle component. The objective is to minimize the weight as well as the cost. The designer wants to have light material, but the marketing manager requires it to be as cheap as possible. Therefore, the material has a constraint of stiffness and yet price.

One of many options of decision support tools is the Multi-Criteria Decision Analysis (MCDA). MCDA is a decision-support tool for exploring issues and making decisions involving multiple criteria. The tool allows technical, economic, environmental, even social criteria, including setting up the priorities. It can be used to generate qualitative data about decision-making criteria, rank decision options, and analyze the background for rank positions. Scores are defined based on references from existing research, guidelines, and standards. Scores are related to general action needed and the risk of accidents. Various criteria are then weighted to reflect the relative importance of different criteria. The weighted sum of the different chosen criteria is used to rank the options.

In pavement design, there is a necessity of having a holistic understanding of the functional performance regardless of various standards and texturing methods available. Thus, the optimum decision can be made to fulfil design requirements. In real life, many decisions involve trade-offs, which is related to conflicting constraints. This is also the case in designing surface pavement to meet the functional performance required.

The surface of pavements has been studied for decades. Many surface treatments have been done by roadway authorities to improve the functional performance of

A. A. Bawono, *Engineered Cementitious Composites for Electrified Roadway in Megacities*, Springer Theses, https://doi.org/10.1007/978-3-030-88542-7_5

the pavements both for concrete and asphalt and: skid resistance, surface water drainage, pavement evenness, noise, and durability. However, there are so many different standards for those criteria applied, which might be causing misinterpretation when decision-makers are about to select the right surface treatment. As an example, is when they need to compare in which texturing methods are applied by different authorities who is using different standards. Firstly, various standards related to the pavement's functional surface performance applied by different authorities. Secondly, there are many different measurement methods for the functional performance of the surface pavement. Thirdly, the surface treatment methods used to improve the surface pavement are varying. Therefore, it is not an easy task to understand, compare one to another surface treatment method with different criteria, and select the optimum method. Yet, fewer studies were found on comprehensive analysis on determining which surface treatment leads to better functional performance.

5.1 Functions, Compound Objectives, and Multiple Constraints Within Functional Performance of Surface Pavements

Following an approach explained by Ashby [1], several functions, compound objectives, and multiple constraints can be found in design requirements on the functional performance of pavements. As functions, there are mainly two criteria related to surface pavements: a design for new roads, and design for retexturing roads. Design for new roads is needed when a new line or roadway is required and decided to be built. The retexturing road is mostly required when the existing roads were built some time ago.

The objectives of road authorities when they are considering such a design (texturing methods) for surface pavement can be different. Typical objectives are elaborated in Fig. 5.1. It should be noted that in many cases, the road authority has more than one objective. As an example, an existing road located in the residential area needs to be treated. The objectives are to reduce the noise as well as to minimize the cost. However, road construction in highway needs more objectives, including quick to be opened for traffic, less noise, high skid resistance, and high comfort.

Ashby [1] explained that even one objective might require more than one constraint. As consequences of different and compound objectives, multiple constraints need to be considered. For example, if the objective is to improve the skid resistance. The constrain will include more parameter, including microtexture, macrotexture, and perhaps local material availability.

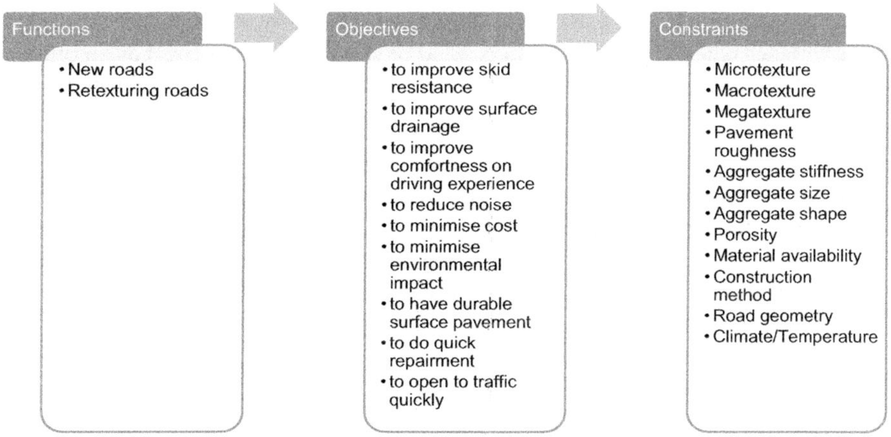

Fig. 5.1 Functions, objectives, and constraints on functional performance

5.2 Scope of MCDA Study

The scope of MCDA needs to be defined as the functional performance can be extremely varied. For example, the design requirement of skid resistance is varying between roadway, roundabout, and pedestrian. Thus, in this study, the scope of MCDA includes:

- Level of the road network: highway or main arterial road,
- New roads or newly retextured,
- Surface texture methods for concrete and asphalt,
- Seven criteria: microtexture, macrotexture, surface drainage, noise, pavement roughness, cost, and durability.

5.3 Methodology of MCDA

This study aims to find an optimum solution for surface pavement from the safety aspect using MCDA. MCDA has many advantages over informal judgement unsupported by analysis, including open and explicit. The choice of objectives and criteria that any decision-making group needs to analyze and change if is inappropriate. Scores and weights, when used, are also explicit and can be developed according to the established techniques.

The first step is data mining of functional performance from texturing methods. Data mining was done by collecting secondary data from previous research published in papers, reports, and any other reliable documents. It is expected to have many missing data of criteria for many texturing methods. Thus, an empirical formula with a strong correlation is used to fulfil the gaps. Then it is statistically analyzed to check whether the quality of the data is good or should be categorized as an outlier.

MCDA techniques commonly apply numerical analysis to a performance matrix in two stages:

1. Scoring: the expected consequences of each option are assigned a numerical score on the strength of a preference scale for each option for each criterion. More preferred options score higher on the scale, and less preferred options score lower. Many studies apply scales extending from 0 to 5, where 0 represents a real or hypothetical least preferred option, and 5 is associated with a real or hypothetical most preferred option. All options considered in the MCDA would then fall between 0 and 5.
2. Weighting: numerical weights are assigned to define, for each criterion, the relative valuations.

A literature review on established guidelines and standards was done to develop the new score. Results have shown that various standards are applied. Therefore, general rules need to be created. Each value of the score will be highly dependent on the general actions needed to improve the functional performance of the pavement. It is also related to the risk of accidents. After the scoring was done, the next step is weighting. Weighting is important as not every criterion has similar importance or urgency level.

In many cases, one criterion can be more important than others, e.g., most authorities prioritize safety more than comfort. With a combination of scoring and weighting, ranking can be done as a final step. The ranking can be used as a reference for a better or worse texturing method for overall criteria. However, in this study, ranking might be not critical, as this study will be focusing on skid resistance and surface water drainage. The flow chart of the methodology is illustrated in Fig. 5.2.

$$\text{Weighted sum} = \sum_{j=1}^{n} w_j x_{ij}, \text{ for } i = 1, 2, 3, \ldots, m_1 \tag{5.1}$$

where

- w relative weight of importance (see Table 5.10)
- x performance value
- n decision criteria (microtexture, macrotexture, pavement roughness, noise, surface drainage, cost, and durability).

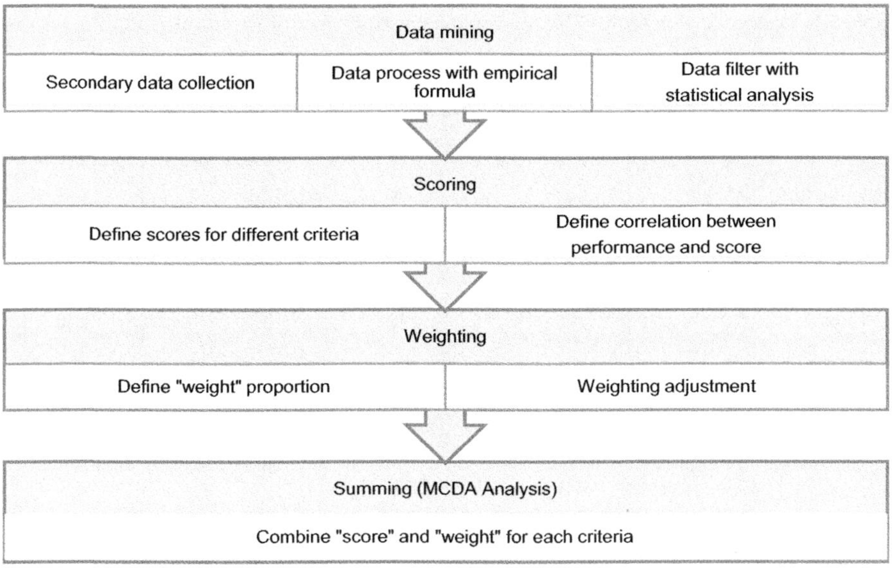

Fig. 5.2 Methodology of MCDA for functional performance of pavement [3]

5.4 Data Mining

5.4.1 Data Collection

More than 10,000 data are collected for different texture methods based on different test standards. Five different criteria of functional performance are included: microtexture, macrotexture, surface pavement drainage, noise, pavement unevenness. Test methods for each criterion included in this MCDA study are summarized in Table 5.1. Additionally, cost and durability, are considered for this MCDA study, as most authorities highly considered these two criteria.

For asphalt pavement (coded as AC), there are about 35 different texturing methods found. Those texturing methods are then labelled with a code, as it will make it easier to be coded and visualized in charts later. On the other side, there are about 37 different texturing methods summarized for concrete pavements (coded as PCC). The texturing method can be categorized into two: New and or overlay (coded as NO) and retexturing or friction restoration (coded as RF). For example, the meaning of the coding is given: ACNO means texturing method for Asphalt Pavement for New and Overlay. Meanwhile, PCCRF means a texturing method for Concrete pavement for Retexturing and or Friction restoration.

Table 5.1 Data collection for MCDA for functional performance of pavement

Criteria	Index	Test 1	Test 2	Test 3	Test 4
Microtexture	Microtexture	BPT (slider)			
	Friction	BPT (slider)	DFT (slider)	DFTester (locked wheel)	SCRIM (side force)
Macrotexture	MPD	Laser	CTM	SPM	
	MTD	Laser	CTM	SPM	
	RMS	CTM			
	ETD	Laser	CTM		
Surface pavement drainage	Water outflow	Outflow meter			
Noise	Wayside (traffic) noise	SPB	CPB	CTIM	
	Tire-pavement noise	OBSI	CPX		
	In-vehicle noise	Microphone			
Pavement roughness	IRI	Laser	Profilometer		
	Profilograph index	Laser	Profilometer		
	Ride Number	Laser	Profilometer		

5.4.2 Data Process with Empirical Formula

Many data are still missing for some criteria from each texturing methods. Thus, there is a need to find a solution on how to fulfil the data. One solution is by using an equation that has a close correlation (R^2) between each criterion. These equations are referred to previous research and mainly are based on empirical studies (Table 5.2).

5.4.2.1 Correlation Between Microtexture/Macrotexture and Friction

The correlation between microtexture and friction has been mentioned by Henry [7]. The equation is given:

$$\mathrm{BPN} = 57.9 \times \mathrm{DFT}(20) + 23.1 \tag{5.2}$$

where

BPN British Pendulum Number
DFT DFTester.

Table 5.2 Correlation between criteria of functional performance of pavement

Correlation	Micro-texture	Macro-texture	Friction	Rough-ness	Surface drainage	Noise
Microtexture			Equation (5.2)			Equation (5.15)
Macrotexture		Equations (5.4, 5.5)	Equations (5.3, 5.6)		Equations 5.7, 5.8	Equations (5.9, 5.10, 5.11, 5.12)
Friction			Equations (5.13, 5.14)			
Roughness				Equation (5.18)		
Surface drainage						
Noise						Equations (5.16, 5.17)

The correlation between macrotexture and friction [7] is given as follows:

$$F(60) = 0.081 + 0.732 \times DFT(20) \times e^{([20-60])/S_p} \quad (5.3)$$

where

F(60) Friction
DFT DFTester
S_p Speed number.

5.4.2.2 Correlation in Macrotexture

The correlation between CT meter MTD and CT meter MPD based on the Wallops equation [7] is shown below:

$$MTD = 0.952 \times MPD + 0.091 \quad (5.4)$$

where

MTD Mean Texture Depth
MPD Mean Profile Depth.

The correlation between MTD from sand patch method and MPD from laser measurement [5] is given as follows:

$$MTD(SPM) = 1.0812 \times MPD(laser) + 0.0697 \quad (5.5)$$

where

MTD Mean Texture Depth
MPD Mean Profile Depth.

5.4.2.3 Correlation Between Macrotexture and Friction

The correlation between macrotexture and friction [7] is given as follows:

$$S_p = 14.2 + 89.7 \times MPD \quad (5.6)$$

where

MTD Mean Texture Depth
MPD Mean Profile Depth.

5.4.2.4 Correlation Between Macrotexture and Surface Drainage

The correlation between MTD and OFT [2] is given as follows:

$$\mathrm{MTD} = \frac{3.114}{\mathrm{OFT}} + 0.636 \tag{5.7}$$

where

MTD Mean Texture Depth (mm)
OFT Water Outflow Time (s).

Another correlation between MTD and OFT [10] is given as follows:

$$\mathrm{OFT} = 2.647 \times \mathrm{MTD}^{-0.789} \tag{5.8}$$

where

MTD Mean Texture Depth (mm)
OFT Water Outflow Time (s).

5.4.2.5 Correlation Between Macrotexture and Noise

The correlation between macrotexture (MPD) and tire-pavement noise ($R^2 = 0.59$) [8] is given as follows:

$$\text{Noise (OBSI)} = 3.13\mathrm{MPD} + 95.57 \tag{5.9}$$

where

OBSI tire-pavement noise (on-board sound intensity)
MPD Mean Profile Depth.

The correlation between macrotexture (MTD) and tire-pavement noise has a high correlation ($R^2 = 0.80$), depends on the speed [11]. The equation is given as follows:

$$\text{Noise (OBSI)} = 1826.2\mathrm{MTD} + 24809 \tag{5.10}$$

$$\text{Noise (OBSI)} = 1404.9\mathrm{MTD} + 17237 \tag{5.11}$$

$$\text{Noise (OBSI)} = 892.67\mathrm{MTD} + 9659.9 \tag{5.12}$$

where

OBSI On-board sound intensity

MPD Mean Profile Depth
MTD Mean Texture Depth.

5.4.2.6 Correlation in Microtexture and Friction

The correlation between British Pendulum Number (BPN) and DFT(20) has a high correlation with $R^2 = 0.86$ [7]. The equation is given as follows:

$$\text{BPN} = 57.9 \times \text{DFT}(20) + 23.1 \tag{5.13}$$

where

BPN British Pendulum Number
DFT DF Tester.

The correlation between F_{60} and FN40S is given as follows:

$$F_{60} = 0,045 + 0,925 \times \text{FN40S} \times e^{5/S_p} \tag{5.14}$$

where

F_{60} Friction
FN40S Friction Number
S_p speed number.

5.4.2.7 Correlation Between Microtexture/Macrotexture and Noise

The correlation between macrotexture and noise in concrete pavement has a strong correlation, $R^2 = 0.61$ [6]. The equation is given as follows:

$$\begin{aligned}\text{Log (SI)} = & -0.35 + 0.37\text{x } TD_{TRAN} + 0.62 \times GTI_{EAC} + 0.20 \times GTI_{LTi} \\ & + 0.01 \times \#\text{Jts} + 0.06 \times \text{Ln(traffic)} + 0.53 \times MPD_{HS}\end{aligned} \tag{5.15}$$

where

TD_{TRAN} Transverse texture direction (1 = transverse, 0 = otherwise)
GTI_{EAC} GTI for Exposed Aggregate Concrete texture (1 = EAC, 0 = otherwise)
GTI_{LTi} GTI for longitudinal-tine texture (1 = longitudinal-tine, 0 = otherwise)
#Jts Number of joints per 305 m
Ln(traffic) Log normal cumulative traffic applications
MPD_{HS} MPD from high-speed profiler (mm).

5.4.2.8 Correlation in Noise

There is a high correlation between the wayside noise and tire-pavement noise [4]. CPB and CPX have a correlation of 79% ($R^2 = 0.79$). CPX and OBSI has a correlation of 93%. The equations are shown below:

$$CPB = 0.8 \times CPX - 2.3 \tag{5.16}$$

where

CPB Controlled pass-by
CPX Close proximity.

$$CPX = 1.04 \times OBSI - 6.52 \tag{5.17}$$

where

CPX Close Proximity
OBSI On-Board Sound Intensity.

5.4.2.9 Correlation in Pavement Roughness

The correlation between PSR and pavement roughness [9] is given as follows:

$$PSR = 5e^{-0.19(IRI)} \tag{5.18}$$

where

PSR present serviceability rating
IRI International roughness index.

5.4.3 Data Filter with Statistical Analysis

It is expected that a lot of data will be collected. Various data from different sources might create big differences (standard deviation). Therefore, this data will be statistically analyzed before it can be processed further. The boxplot is used to display the distribution of data based on minimum, first quartile, median, third quartile, interquartile range, and maximum.

5.5 Scoring on Criteria for Functional Performance

After the data process, the next step is the scoring. Scoring criteria is set up based on the literature review from guidelines and standards issued by roadway authorities. Then, these scoring criteria will be used as a reference for categorizing functional performance.

5.5.1 *Scoring Criteria*

Scoring criteria is set up based on the literature review from guidelines, standards, and related documents issued by roadway authorities. The criteria are set from 1 to 5 (1 = lowest performance, and 5 = highest performance), and they are mainly based on the performance, required actions, and risk of accidents. Statistical analysis is used only when no similar scoring criteria are found on existing standards, guidelines, and related documents issued by authorities. The scoring criteria are summarized in Table 5.3.

5.5.2 *Scoring on Microtexture*

The proposed scoring for microtexture (friction) according to a few existing standards and guidelines is summarized below (Table 5.4).

Table 5.3 Scoring category and their criteria for MCDA

Category	Scoring	Standard related	Typical action	Risk to accidents	Statistical analysis
Excellent	5	Far beyond standard	No further action required	No	Above Q3
Good	4	Slightly above standard	Regular maintenance	No	Median to Q3
Fair	3	Meet the standard	Regular maintenance	No	Standard to median
Poor	2	Slightly below Standard	Maintain monitoring, data collection, and take corrective action as required	**Yes**	
Very poor	1	Far below standard	Rehabilitation or reconstruction is needed	**Yes**	

Table 5.4 Scoring for microtexture

Score (category)	BPT	DF tester		Locked wheel	
	Slider	DFT(20)	F(60)	FN40R	FN40S2
	BPN	–	–	–	–
5 (Excellent)	>75	>90	>40	>50	>50
4 (Good)	65–75	70–90	35–40	45–50	40–50
3 (Fair)	55–65	50–70	28–35	35–45	28–40
2 (Poor)	45–55	40–50	24–28	26–35	20–28
1 (Very Poor)	<45	<40	<24	<26	<20

Table 5.5 Scoring for macrotexture

Score (category)	Laser		CT meter		Sand patch
	MPD	MTD	MPD	MTD	MTD
	mm	mm	mm	mm	mm
5 (Excellent)	>1.15	>1.2	>1.15	>1.2	>1.2
4 (Good)	0.9–1.15	1–1.2	0.9–1.15	1.0–1.2	1.0–1.2
3 (Fair)	0.8–0.9	0.8–1.0	0.8–0.9	0.8–1.0	0.8–1.0
2 (Poor)	0.35–0.8	0.4–0.8	0.35–0.8	0.4–0.8	0.4–0.8
1 (Very poor)	<0.35	<0.4	<0.35	<0.4	<0.4

5.5.3 Scoring on Macrotexture

The proposed scoring for macrotexture according to a few existing standards and guidelines is summarized below (Table 5.5).

5.5.4 Scoring on Pavement Surface Pavement Drainage

The proposed scoring for surface pavement drainage according to a few existing standards and guidelines is summarized below (Table 5.6).

5.5.5 Scoring on Noise Reduction

The proposed scoring for noise level according to a few existing standards and guidelines is summarized below (Table 5.7).

Table 5.6 Scoring for pavement drainage

Score (category)	OFM
	OFT
	sec
5 (Excellent)	<2
4 (Good)	2–5
3 (Fair)	5–10
2 (Poor)	10–50
1 (Very poor)	>50

Table 5.7 Scoring for noise level

Score (category)	Noise level	
	Tire-pavement	
	OBSI	OBSI
	dB (A)	dB (A)
5 (Excellent)	<102	<102
4 (Good)	102–104	102–104
3 (Fair)	104–106	104–106
2 (Poor)	106–108	106–108
1 (Very poor)	>108	>108

5.5.6 Scoring on Pavement Roughness

There are many methods used to measure pavement roughness according to a few existing standards and guidelines. This study includes three methods: Profilograph Index, International Roughness Index (IRI), and Pavement Serviceability Ratio (PSR). The proposed scoring for pavement roughness is summarized below (Table 5.8).

Table 5.8 Scoring for pavement roughness

Score (category)	Pavement roughness			
	PrI	IRI		PSR
	m/km	mm/m	In/mi	–
5 (Excellent)	<110	<0.95	<60	>4.0
4 (Good)	110–160	0.95–1.48	60–94	3.5–4.0
3 (Fair)	160–230	1.48–1.88	95–119	3.0–3.5
2 (Poor)	230–450	1.88–2.68	120–170	2.5–3.0
1 (Very poor)	>450	>2.68	>170	<2.5

5.5.7 *Scoring on Cost and Durability*

The proposed scoring for the cost of surface treatment and durability according to a few existing standards and guidelines are summarized Table 5.9. The cost presented in three different currencies. The conversion rates used are based on the date of 29 October 2018 issued by Bloomberg (www.bloomberg.com), which are: 1.3826 for USD/SGD and 0.8792 for USD/Euro. The durability presented in years (Table 5.9).

5.6 Weighting on Criteria for Functional Performance

The next step of MCDA is weighting. Weighting is a crucial consideration as not every criterion has similar importance or urgency level. One criterion can be more important than others, e.g., most authorities prioritize safety more than comfort. The weight proportion used in this study is based on a previous study [7]. Their study summarized that safety has more prioritization than other criteria (refer to Table 4.1). However, a few adjustments were made for this MCDA, including:

- The weight of durability in this MCDA study is 1.2, which is the average durability for the USA and other countries.
- There is no criterion of cost from the NHCRP reference. However, it is mentioned that durability is highly important as it is related to budgeting consideration for the authority. Therefore, in this study, MCDA is considered as important as durability, which is 1.2.
- The weight for both microtexture and macrotexture in MCDA is similar to the weight of skid resistance from the NCHRP study, as both microtexture and macrotexture are a detailed factor of skid resistance. However, the weight is averaged.
- The weight of surface drainage in this MCDA is the average weight for splash and spray for the USA and other countries.
- The weight of tire-pavement noise in this MCDA is the average weight for exterior noise for the USA and other countries.

Table 5.9 Scoring for cost and durability

Score (category)	Cost			Durability
	USD $/m^2	Euro €/m^2	SGD $/m^2	Years
5 (Excellent)	<1.06	<0.93	<1.47	> 11
4 (Good)	1.06–1.98	0.93–1.74	1.47–2.74	9.5–11
3 (Fair)	1.98–3.86	1.74–3.39	2.74–5.34	7.5–9.5
2 (Poor)	3.86–10.60	3.39–9.32	5.34–14.66	4.6–7.5
1 (Very poor)	>10.60	>9.32	>14.66	<4.6

Table 5.10 Weight proportion for MCDA

Design criteria	Reference ranking		Proposed in MCDA study	
	USA	Other countries	Weight proportion	Inversed weight (ratio to 1)
Durability	1.1	1.3	1.2	0.18
Cost			1.2	0.18
Skid resistance	1.2	1.4		
Microtexture			1.3	0.17
Macrotexture			1.3	0.17
Splash and spray	2.0	1.8		
Surface drainage			1.9	0.13
Exterior noise	2.4	2.2		
Tire-pavement noise			2.3	0.10
In-vehicle noise	2.4	2.4		
Rolling resistance	2.7	2.7		
Tire wear	2.7	2.9		
Pavement roughness			2.7	0.06

- The weight of pavement roughness in this MCDA is the average of weight rolling resistance and tire wear. The pavement roughness considered has a similar weight as it is closely affected by rolling resistance and tire wear.

The next step is to inversely calculate the weight from the "3 to 1" scale into the "1 to 5" scale, with 1 is the lowest priority, and 5 is the highest priority. Then, the last step is to convert all of the weight into a ratio to 1, which means the sum of weight for all seven criteria is 1. The summary of the previous study, the new weight proportion for the MCDA study, and the inversed ratio for related weight are summarized in Table 5.10.

5.7 Results

More than 10,000 data from different texturing methods for asphalt and concrete pavements have been collected and analyzed for seven different criteria (constraint) of the functional perfomance: microtexture and friction, macrotexture, water outflow, pavement roughness, and tire-pavement noise, cost, and durability.

According to the MCDA results, best practice of texturing methods for skid resistance and macrotexture include longitudinal diamond grooving, transverse diamond grooving, and turf drag. Additionally, exposed aggregate, longitudinal tining, and burlap drag can be used to improve skid resistance to the required safety standards. In terms of surface drainage, diamond grooving gives the best results. However, other texturing methods such as diamond grinding, tining, burlap drag, exposed aggregate resulting in a satisfactory performance for water outflow. The optimum texturing methods for both low tire-pavement noise and skid resistance include longitudinal tining, turf drag, broom drag, exposed aggregate, and followed by diamond grinding. The least solution for noise is transverse diamond grooving.

Texturing methods for concrete pavements were checked closely, as concrete pavement considered closely related to PUTW, which both categorized as a cementitious material. Thus, typical texturing methods such as tining, grinding, and grooving can also be implemented in PUTW. Although texturing methods for asphalt pavement can also be considered if it is applied as a surfacing layer. However, the surfacing layer consequently adding more work, and cost.

5.7.1 Radar Chart Analysis

The first analysis is done with the radar chart to evaluate each texturing method. Radar chart, also known as spider chart, web chart, polar chart, or star plot, is a two-dimensional chart to display multi-dimensional data structure. This spider chart is useful as it can be used to compare the average score of different characteristics of constraints (in this study includes: skid resistance, surface water drainage, tire-pavement noise, pavement roughness, durability, and cost). Further, the radar chart gives an overall picture of how each texturing method performs, as it gives the impression on how good or bad for each constraint. The weakness of this radar chart is, that it gives no clarity when the chart is used to compare different texturing methods. For comparison, a different type of analysis is needed, which is described in the next sub-chapters.

The functional performance of each texturing method for asphalt and concrete pavements are investigated with radar chart analysis. Seven constraints (microtexture, macrotexture, pavement roughness, surface water drainage, tire-pavement noise, durability, and cost) are included in each texturing method which the value is scaled from 0 to 5. More results are elaborated in Appendix C.

5.7.2 Multiple Two-Constraints Analysis on Functional Performance

The second analysis is done by plotting multiple constraints scenario. This analysis allows a comparison of many texturing methods. However, it is limited to two constraints. The plot is not determined to give a correlation between constraint (criterion). Instead, it gives a picture of what is the performance of two constraints for each texturing method.

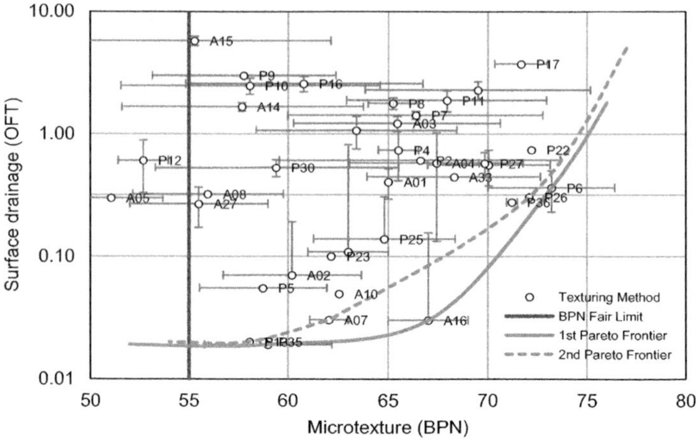

Fig. 5.3 Two-constraints analysis

Figure 5.3 gives an example of two-constraints combination between microtexture (BPN) and surface drainage (OFT, seconds). Within each graph, first and second frontier is included which gives identification for the first and the second-best options for those two constraints. However, it should be noted that they are not to be outside boundaries set by two lines for each constraint. As the boundary line have been set according to scoring criteria. It means any value outside the boundary lines gives bad results (Poor and Very Poor performance). More results for combination on macrotexture, microtexture, tire-pavement noise, pavement roughness, durability, and cost are elaborated in Appendix C.

5.7.3 Multi Criteria Decision Analysis

The last analysis method is done by using MCDA. Every texturing method has been scored and weighted. As the final step, each value from seven constraints is summed. As results, each texturing method has a total value. The total value can be used as criteria for ranking the methods. The maximum value is 5. The weighted sum for all texturing methods for asphalt pavements and concrete pavements are summarized in Figs. 5.4 and 5.5, respectively.

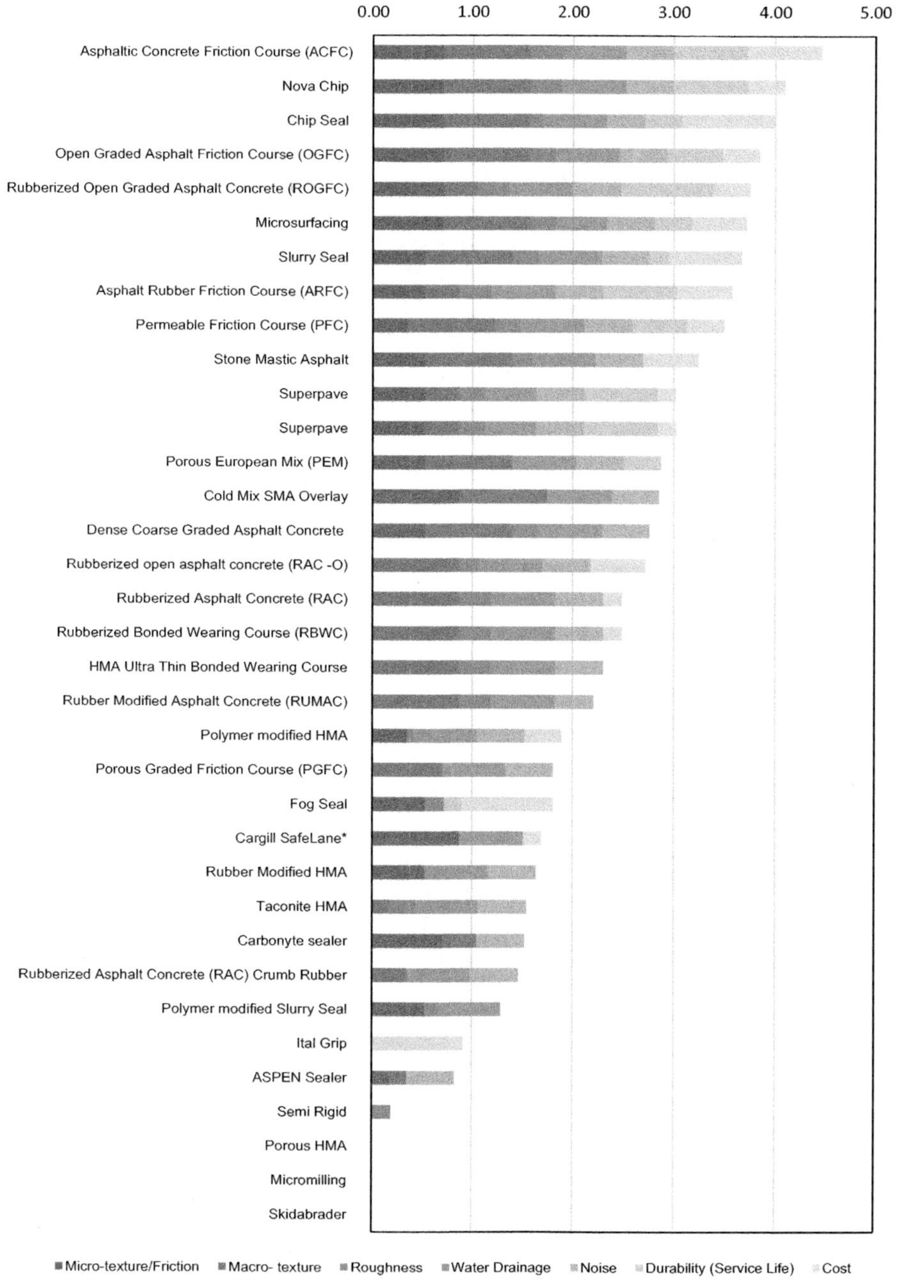

Fig. 5.4 Weighted sum of texturing methods for asphalt pavement

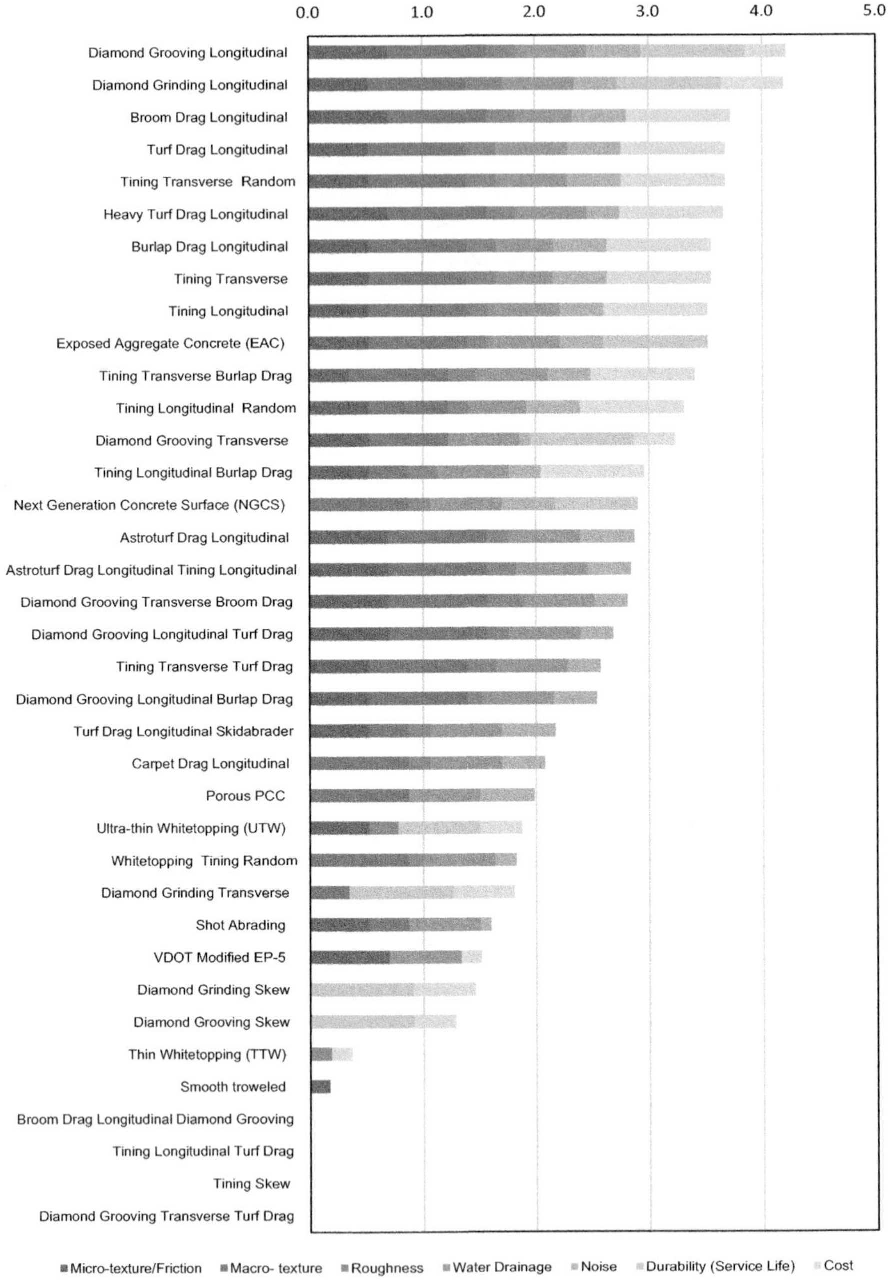

Fig. 5.5 Weighted sum of texturing methods for concrete pavement

References

1. Ashby MF (1999) Materials selection in mechanical design. Butterworth-Heinemann, Oxford
2. ASTM E2380M-15 (2015) Test method for measuring pavement texture drainage using an outflow meter. ASTM International, West Conshohocken, PA
3. Bawono AA, Caliskan E, Lechner B, Yang E-H (2019) Multi criteria analysis for the functional performance of surface pavement. In: Advances in engineering materials, structures and systems: innovations, mechanics and applications, Cape Town, South Africa, August. CRC Press, London
4. Donavan PR (2006) Comparative measurements on tire pavement noise in Europe and the USA. FHWA/CA/MI-2006/09, California
5. Hall J, Smith K, Titus-Glover L, Wambold J, Yager T, Rado Z (2009) Guide for pavement friction. Washington, DC
6. Hall J, Smith K, Littleton P (2009) Texturing of concrete pavements. NCHRP report 634, Washington, DC
7. Henry J (2000) Evaluation of pavement friction characteristics: a synthesis of highway practice, Washington, DC
8. Hong SJ, Park S-W, Lee SW (2018) Tire-pavement noise prediction using asphalt pavement texture. KSCE J Civ Eng 22(9):3358–3362. https://doi.org/10.1007/s12205-018-9501-3
9. Paterson W (1986) International Roughness Index relationship to other measures of roughness and riding quality. Transport Res Record 1084
10. Sarsam SI (2012) Field evaluation of asphalt concrete pavement surface texture and skid characteristics. 5th Eurasphalt & Eurobitume Congress, Istanbul, 13–15th June 2012
11. Saykin VV (2011) Pavement macrotexture monitoring through sound generated by the tire-pavement interaction. Master of Science in Civil Engineering, Northeastern University

Chapter 6
Study of Mechanical Properties

This chapter describes the results of the mechanical properties study completed in this thesis. The main objective of the study is to investigate the functional performance of pavement made of ECC material. However, the study and understanding of the ECC characteristics are required before doing any modification for the existing mixture design. This chapter also explains the mechanical properties due to the modification of the mixture design of ECC and, if any, possible effect to the structural performance of the applied pavement.

The non-existence of the aggregates is results in low skid resistance for ECC, as there are less microtexture and macrotexture. The development of ECC design is ideally based on the performance-driven design approach (PDDA), which offers a systematic engineering solution rather than trial-and-error empirical experiments [5]. However, in this study, the methodology has been modified since the functional performance has to be also fulfilled by improving the microtexture and macrotexture. Thus, aggregates need to be included in the modified mixture design. Consequently, there will be an adjustment needed on the micromechanics model, microstructure tailoring, and processing.

6.1 Experimental Program

6.1.1 Materials and Mixture Design

Two types of modified ECC have been developed in this study. The first one is the modified ECC mixed with corundum, namely ECC-Cor. And the second one is the modified ECC mixed with silica sand, named ECC-SS. ECC-Cor is using Type I Portland cement, class F fly ash (FA), micro-silica sand (mean and maximum grain size of 110 μm and 200 μm, respectively), polyvinyl alcohol (PVA) fibers (a length of 12 mm, a diameter of 39 μm, a nominal tensile strength of 1600 MPa), and a high-range water reducing (HRWR). Corundum with a maximum particle

A. A. Bawono, *Engineered Cementitious Composites for Electrified Roadway in Megacities*, Springer Theses, https://doi.org/10.1007/978-3-030-88542-7_6

Table 6.1 Mixture design of ECC-Cor (kg/m^3)

Mixture	Cement	Fly ash	Sand	Water	Super pplasticizer	PVA fiber	Corundum
ECC-Cor 0	559	671	447	327	2.3	26	0
ECC-Cor 50	559	671	224	327	2.3	26	224
ECC-Cor 60	559	671	179	327	2.3	26	268
ECC-Cor 70	559	671	134	327	2.3	26	313
ECC-Cor 80	559	671	89	327	2.3	26	358
ECC-Cor 90	559	671	45	327	2.3	26	402
ECC-Cor 100	559	671	0	327	2.3	26	447

size of 2.38 mm was used as fine aggregates for improving the surface textures of ECC. Table 6.1 summarizes the mixture design of ECC, incorporating a different amount of corundum (ECC-Cor). A control ECC mixture without the addition of corundum (ECC-Cor 0) was also prepared. All ECC-Cor mixtures have a low water-to-cementitious material ratio (w/cm) of 0.27. Micro-silica sand was replaced with corundum at different dosage, ranging from 50–100%, and the mixture was denoted as 'ECC-Cor #', indicating # with % of micro-silica sand was replaced with corundum.

The second mixture is ECC-SS. Materials used include Type I white Portland cement, water, Polyester fibers (PE) 19 mm in length, oil coating: 1.2%, polycarboxylic-ether type high-range water-reducing admixture plasticizer (ADVA 181), viscosity modifying admixtures (VMA) with a type of Methyl Cellulose and aggregate silica sand with hardness is 7 Mohs' scale, a density of 2.65 g cm^3, with a maximum particle size of 2.38 mm as suitable aggregates. Table 6.2 summarizes the mixture design of ECC-SS.

For full-scale, four 1000L-plus trial mixing were carried out. The mixture proportion of the main raw materials, dimension of cast slab, slab surface treatment can be found in Table 6.3.

Table 6.2 Mixture design of ECC-SS (kg/m^3)

Mixture	White cement	Water	Super plasticizer	PE fiber (% by vol.)	VMA	Silica sand
ECC-SS 00	1474	516	2.5	1.8	1.50	0
ECC-SS 01	1474	516	2.5	1.8	1.50	147
ECC-SS 02	1474	516	2.5	1.8	1.50	295
ECC-SS 03	1474	516	4.0	1.8	1.50	442

Table 6.3 Mixture of the additional four large-scale mixing trials

No	Mix	Mix proportions (kg/m^3 or g/L)						V (L)	Fiber dispersion and workability	Dimension of cast slabs
		OPC	FA	SS	Water	Fiber	Cor			
1	BC^	727	484	363	363	26	0	1244	Poor	dumped
	CBC^	755	482	115	348	15	303	150	Poor	
2	BC	488	780	432	291	28	0	1037	a bit bleeding	3.6 m × 2.4 m × 0.07 m
	CBC	415	771	119	308	18	308	150	Good	
3	BC	504	756	454	291	26	0	1031	Good	Cast-in-site
	CBC	522	783	130	261	20	339	150	Good	
4	BC	620	676	468	279	26	0	1006	Good	3 of 0.9 m × 1.2 m × 0.07 m
	CBC	415	771	119	250	18	334	150	Good	

6.1.2 Specimen Preparation

Each of ECC-Cor and ECC-SS mixture was prepared in a planetary mixer. Cement, fly ash, and micro-silica sand are dried mixed first. Water was then added, followed by the addition of superplasticizer and mixed for 3–5 min until the mix achieved the required consistency. Lastly, the PVA fiber and aggregates (corundum or silica sand) were added and mixed for three to five minutes. The fresh mixture was cast into a prism (280 mm × 70 mm × 50 mm) and plate (500 mm × 300 mm × 10 mm) moulds catered for various types of tests. Specimens were removed from the mould after one day and were cured at normal laboratory air (25 ± 3 °C, 70%RH) for another 27 days before further testing. At the pre-determined age, either one or three days, the top surface of plate specimens was treated with a steel brush to expose the aggregates (corundum or silica sand) on the surface.

6.1.3 Test Program

The compressive test was carried out following ASTM C109/109 M to determine the compressive strength of the cube specimens at the age of 28 days. The test was done under load with a loading rate of 25 kN/min. At least four specimens were tested for each mix. A uniaxial test was carried out to determine the tensile behaviour of ECC-TiO_2 specimens. The coupon specimens were used. At least three specimens were tested for each mix. Tests were conducted with the MTS machine loaded with 25 KN with a loading rate of 0.0025 mm/s. It is equipped with two external linear variable displacement transducers (LVDTs) to measure the displacement. Four-point bending tests were done in accordance with ASTM C78/78 M were carried out on prism specimen to evaluate the flexural strength of specimens. The loading span (80 mm) is one-third of the support span (240 mm), and the rate of deflection at the loading points was controlled at 0.3 mm/min. At least three specimens were tested for each mix. Modulus of rupture (MOR) was calculated, and the ductility was estimated by means of a reverse calculation method.

The measurement of the real contact stresses gathered from the Tekscan I-Scan device. Various sets of loading and inflation pressure are applied to the apparatus. The vertical load is exerted on a single radial tire (Continental 315/80R22.5). Since only static loading is considered, only vertical contact stresses are acquired. Then, the measured data is analyzed and implemented into the verified 3D Finite Element (FE) model. Hence, the ANSYS is employed for numerical analysis. All contact stresses data is successfully imported into ANSYS [2]. The tire imprint in the FE model is totally identical to that from the measuring device. The same model used to study the effects of groove on the structural performance for pavement made of ECC.

The Wheel Tracking Slope (WTS) test was carried out to evaluate the long-term durability of the surface skid resistance. The test was done by following guidelines stated on TRL Report 176. The wheel tracking machine consists of a rubber wheel, which rotates at 42 cycles per minute while applying a load of approximately 520 N. The load traffic simulations run up to 100,000 wheel-passes, which is equivalent to about 3.7 million equivalent single axle loads (ESALS) according to Nichols [6]. During the traffic simulation, the tests were stopped in between to measure BPN and MTD by means of BPT and SPM, respectively. Two mixture formula, i.e., ECC-Cor 50 and ECC-Cor 100, were tested.

6.2 Results

This chapter gives an understanding of the characteristics of the modified ECC mixed with aggregates. Ideally, both structural and functional performance have to fulfil the requirements. Modification on the PDDA approach is needed as the original approach is not considering the functional performance in the desired mechanical properties, which are skid resistance and surface water drainage. Through some tests, it is proven that some modification on the processing and microstructure tailoring process allows the achievement of the desired mechanical properties. The modification includes the properties of aggregates, fibers, other materials, and the processing way. In summary, the inclusion of aggregates, which is needed to improve the microtexture and macrotexture of ECC, does not negatively affect the mechanical behavior of ECC. Similar findings were also reported [7, 9].

6.2.1 *Effects of Aggregates on the Mechanical Behavior*

This study investigates the effects of aggregates on the mechanical behavior of the modified ECC. Table 6.4 summarizes the mechanical properties of the seven ECC-Cor mixes. The first cracking strength and flexural strength were obtained directly from the four-point bending test results, while tensile strain capacity was inversely

Table 6.4 Mechanical properties of ECC-Cor

Specimen	First cracking strength (MPa)	Flexural strength (MPa)	Tensile strain capacity (%)
ECC-Cor 0	3.4 ± 0.4	11.1 ± 0.6	2.1 ± 1.0
ECC-Cor 50	6.0 ± 0.1	11.2 ± 1.1	0.8 ± 0.1
ECC-Cor 60	6.2 ± 0.2	12.1 ± 1.0	1.1 ± 0.2
ECC-Cor 70	6.0 ± 0.4	10.1 ± 1.2	1.1 ± 0.3
ECC-Cor 80	5.9 ± 0.2	11.3 ± 2.2	1.0 ± 0.3
ECC-Cor 90	5.6 ± 2.9	10.9 ± 1.7	1.0 ± 0.4
ECC-Cor 100	5.4 ± 1.2	10.7 ± 1.4	0.8 ± 0.1

Table 6.5 Mechanical properties of ECC-SS

Specimen	Compressive strength (MPa)	Flexural strength (MPa)	Tensile strain capacity (%)
ECC-SS 00	38.06 ± 0.06	9.13 ± 0.56	3.71 ± 1.50
ECC-SS 01	38.56 ± 1.00	9.01 ± 0.24	2.80 ± 0.52
ECC-SS 02	49.02 ± 1.74	10.20 ± 1.46	2.35 ± 0.10
ECC-SS 03	49.10 ± 2.86	4.21 ± 0.01	1.40 ± 0.34

calculated from the load–deflection curve in accordance with ASTM C78/C78M [1]. The first cracking strength of ECC increases while the tensile strain capacity decreases when corundum is incorporated into the mix. This is attributed to the introduction of corundum with a high hardness which increases the matrix toughness and thus reduces the strain hardening and multiple cracking potentials [3]. Nevertheless, all ECC-Cor mixes still possess an adequate tensile strain capacity of around 1%, which is fulfilled the performance requirement for pavement application. Furthermore, the introduction of corundum does not seem to a have negative impact on the flexural strength of ECC-Cor as all mixes maintain high flexural strength between 10 and 12 MPa.

The workability was qualitatively measured, and it can be concluded that the workability of ECC is improved when corundum aggregate is added. This is because the existence of corundum creates more surface area, resulting in less demand for water to wet the surface. Thus, the addition of the aggregates gives higher workability for similar water content. However, there are limits on aggregate particles size as it is affecting fiber dispersion, workability, and matrix toughness which may negatively affect the mechanical performance characteristics of ECC [7].

Table 6.5 summarizes the mechanical properties of the four ECC-SS mixes. The addition of fine size aggregate of silica sand does not significantly affect the mechanical properties of the modified ECC. All parameters, including compressive strength, flexural stress, and tensile strain capacity, are still in the performance range of conventional ECC [4], which are 20 to 95 MPa for compressive strength, 10 to 30 MPa for flexural strength, and 1 to 8% of tensile strain capacity.

In summary, the inclusion of aggregates into the ECC mixture does not negatively influence the mechanical behavior in a significant way. The increase of the volume fraction of the aggregates slightly decreases the flexural strength and tensile performance. On the contrary, the increase of the volume fraction of aggregates increases the compressive strength.

6.2.2 Effects of Aggregates on the First Crack Strength

Aggregates are affecting the first crack strength of the modified ECC. As can be seen in Table 6.4, the first crack strength of modified ECC mixed with aggregates is generally higher than normal ECC. This is due to more energy is needed to initiate

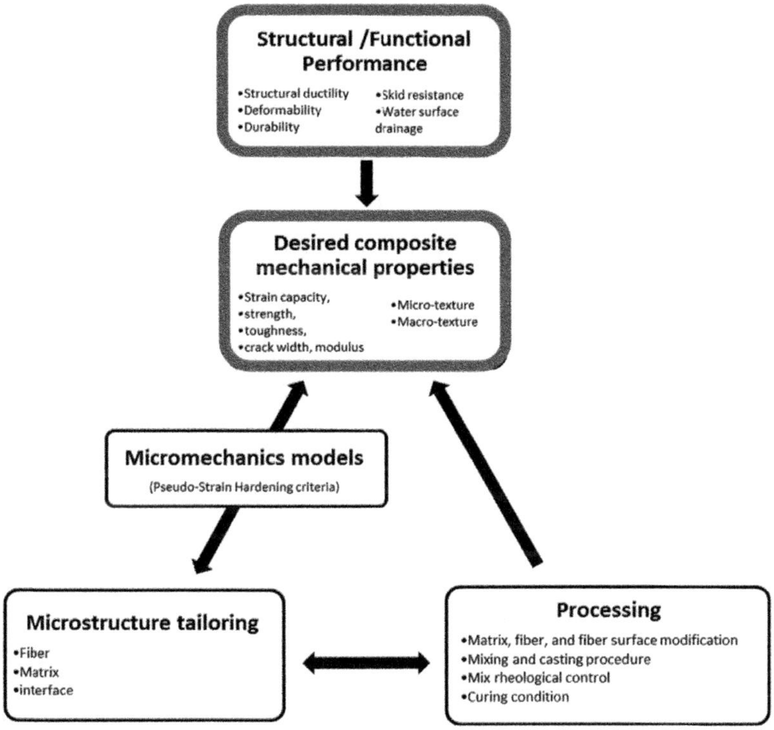

Fig. 6.1 Modified PDDA to improve the functional performance of ECC material

the crack when aggregates are embedded inside the mixture. The illustration is given in Fig. 6.2.

6.2.3 Effects of Fibers to the Mechanical Behavior

ECC with PVA fiber produces good tensile behavior. PVA fiber results in a high interfacial chemical bond, frictional stress, and slip-hardening in a cementitious matrix. The test conducted shows robust pseudo-strain-hardening behavior with the ultimate strain exceeding 2%. It was achieved at a moderate fiber volume fraction of 2%.

Surface coating by oil on fiber is affecting the strain hardening behavior. As the oiling content increases, both frictional stress and interface fracture energy will decrease significantly. As results, the complementary energy of the fiber bridging stress increases with oil content given [10]. Test results have shown that ECC with PVA gives good results in compressive and flexural strengths, which is in line with previous research [8].

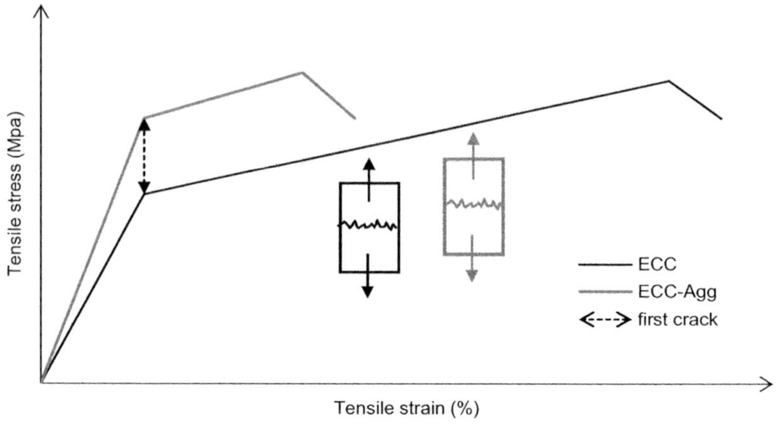

Fig. 6.2 Effects of aggregates on the first crack strength

6.2.4 Effects of Grooves to the Structural Performance of the Slab

The effect of grooves on the PUTW slab's structural performance is investigated, as the thickness design is only 70 mm. It is significantly different from a conventional concrete slab, where commonly, the concrete slab has a thickness of at least 250 to 300 mm.

Measurement of the contact pressures was done with the pressure mapping system device, called Tekscan I-Scan. It has a sensor type of 7101 comprised of sensing elements arranged in 88 rows and 96 columns. The methodology used is based on a previous study [2]. The data collected was then processed using a numerical computing language tool. The magnitude of the contact stresses and their positions are located precisely based on the coordinate system. Different combinations of tire loading (10–50 kN) and tire inflation (600–900 kPa) were used.

The results from the Tekscan I-Scan analysis has shown that there is no negative impact by introducing grooves to the thin PUTW slab. The stresses produced are causing neither cracks nor failure for the slab. The results are shown in Fig. 6.3.

6.2.5 Large Scale Mix

Results from larger-scale mix have shown that there is no significant difference compared to laboratory scale. However, the temperature is one of the main factors that should be considered as a higher temperature leads to a faster curing process, which affects the casting and finishing process. Several important know-hows were learned from the additional four times of large-scale mixing: the quality of the fly ash is indeed very important for desirable fiber dispersion and workability; additional

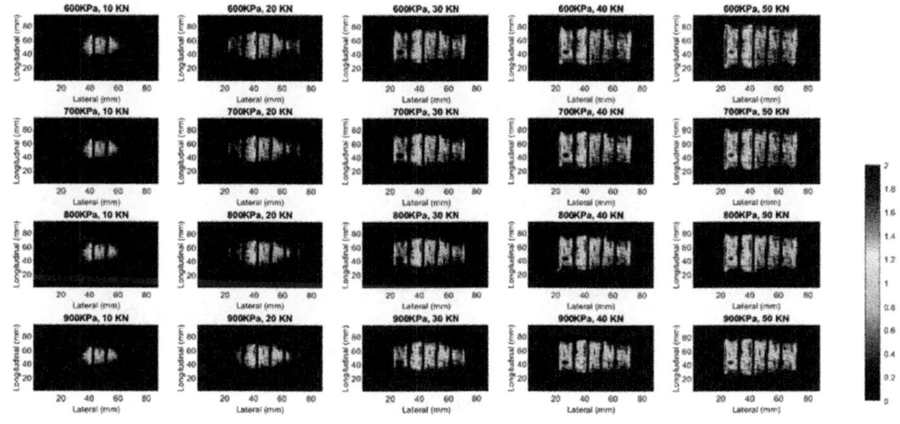

Fig. 6.3 Tekscan results for stresses on grooved ECC slab

Table 6.6 Mechanical properties of ECC-Cor from large-scale mixing

Specimen	Compressive strength (MPa)	Flexural strength (MPa)	Tensile strain capacity (%)
ECC-Cor	57.9 ± 5.1	9.9 ± 0.70	1.90

water (about 1–3% of the binder, i.e., cement plus fly ash) is needed under relatively dry weather; relatively good fiber dispersion may be maintained even with slightly over-dosage of the fibers. The mechanical properties of the mix were characterized by compression, tensile, and flexural test after air curing of 28 days. The results were summarized in Table 6.6. The tensile stress–strain curve of the large-scale mixing specimen was compared with that of the lab-made specimens. The mechanical properties are comparable to the lab version and higher, leading to the conclusion that the technology of large-scaling mixing is mature and ready for field application.

6.2.6 *Durability of Surface Texture*

Figure 6.4 shows the BPN and MTD of ECC-Cor 50 and 100 after different load cycles. As can be seen, the trend shows BPN reduces while MTD remains unchanged with increasing load cycles. Even after 100,000 load cycles, the BPN is still above 65 BPN, and the MTD maintains at 0.34 mm. It can be concluded the surface texture and skid resistance made of ECC-Cor can last until the end of the pavement lifetime without any further intervention. This is mainly due to the hardness of corundum which makes the surface texture of ECC-Cor durable.

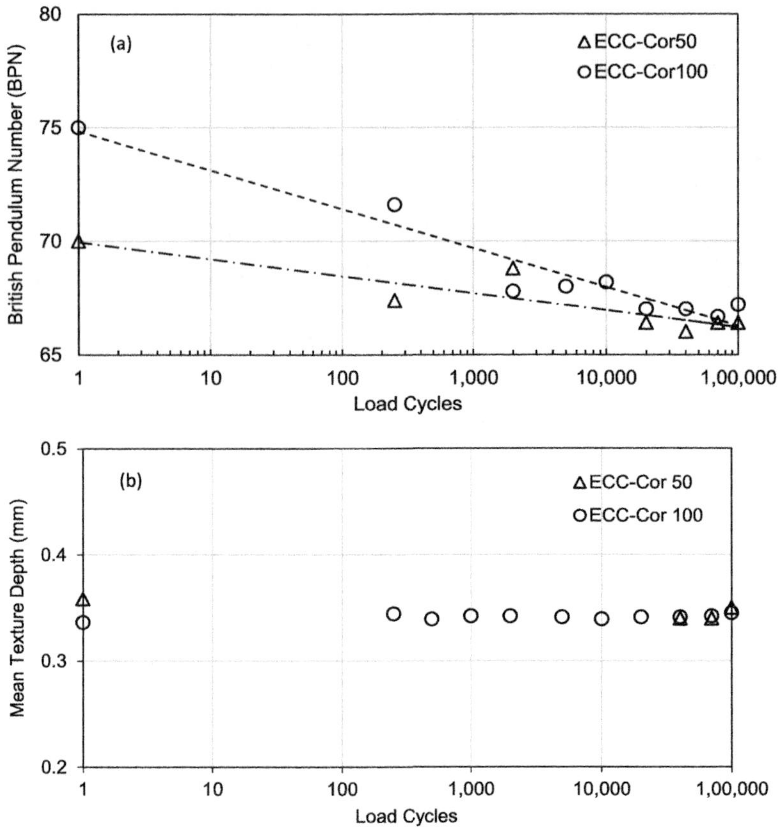

Fig. 6.4 **a** BPN and **b** MTD of ECC-Cor 50 and 100 as a function of load cycles

References

1. ASTM C109 (2020) Standard test method for compressive strength of hydraulic cement mortars. ASTM International, West Conshohocken, PA. https://doi.org/10.1520/C0078_C0078M-18
2. Bawono AA, Lechner B, Alhaz FI (2018) Asphalt pavement analysis with 3D FEM Model using actual tire pavement contact stresses measure with pressure mapping system device. In: Transportation Research Board 97th Annual Meeting, Washington DC, United States
3. Li J, Yang E-H (2017) Macroscopic and microstructural properties of engineered cementitious composites incorporating recycled concrete fines. Cement Concr Compos 78:33–42. https://doi.org/10.1016/j.cemconcomp.2016.12.013
4. Li VC (2007) Concrete construction engineering handbook: engineered cementitious composites (ECC)—material, structural, and durability performance. CRC Press, Michigan
5. Li VC, Wang S, Wu C (2001) Tensile strain-hardening behavior of Polyvinyl Alcohol Engineered Cementitious Composite (PVA-ECC). ACI Mater J 98(6):483–492
6. Nicholls J (1997) Laboratory tests on high-friction surfaces for highways: TRL Report 176, Crowthorne, Berkshire

7. Sahmaran M, Lachemi M, Hossain KM, Ranade R, Li V (2009) Influence of aggregate type and size on ductility and influence of aggregate type and size on ductility and mechanical properties of engineered cementitious composites. ACI Mater J 106(3):308–316
8. Sathishkumar P, Sampathkumar P, Karthik M, Vignesh C (2016) Significance of various fibres on engineered cementitious concrete. Int J Sci Eng Technol 4(1):284–290
9. Ueda N, Kawamoto A (2017) Influence of coarse aggregate on the mechanical properties. Strain-hardening cement-based composites. In: 4th international RILEM conference, Dresden, Germany. Springer
10. Wang S, Li VC (2006) Polyvinyl Alcohol fiber reinforced ECC: Material design and perfomances. In: International RILEM workshop on HPFRCC in structural applications, Michigan

Chapter 7
Study of Skid Resistance: Laboratory Test

7.1 Introduction

This chapter presents the experimental research of the skid resistance for ECC material used for pavement. Data have been collected from various laboratory tests. The skid resistance tests include British Pendulum Test (BPT) for microtexture and the Sand Patch Method Test (SPM) for macrotexture. Different surface treatment methods were evaluated to find the optimum solution to produce good surface properties. Laboratory results show that the skid resistance of the surface pavement can be improved by adopting hard fine aggregates into ECC mix, as it provides both the microtexture and macrotexture. It is also shown that methods on laboratory scale tests can be further applied to full-scale application with results of a surface with good skid resistance.

7.2 Laboratory Test Program

The laboratory test is mainly designed to investigate whether ECC can be modified to improve the skid resistance for pavement application. The first part, which was to maintain the mechanical behavior was discussed in the previous chapter. Thus, this part mainly discusses the test and results related to skid resistance.

A normal ECC with neither aggregates nor surface treatment was investigated. As ECC does not use any aggregates, it is expected to have low results in terms of skid resistance. Therefore, the results will be used as a reference. Two primary types of mixture designs were designed and tested in this study. ECC mixed with corundum aggregate (ECC-Cor) and ECC mixed with silica sand (ECC-SS). The first mixture, ECC-Cor, was designed to achieve high skid resistance. Therefore, corundum, which has high hardness and abrasiveness, is used in the mixture. The second mixture design, ECC-SS, has a slightly different objective. It is to find a mixture design that offers good skid resistance and a brighter surface layer, which

A. A. Bawono, *Engineered Cementitious Composites for Electrified Roadway in Megacities*, Springer Theses, https://doi.org/10.1007/978-3-030-88542-7_7

has less impact on Urban Heat Island (UHI) and better visualization. Thus, silica sand, which has a high hardness (slightly below corundum) and bright color, is used. Different surfacing methods were investigated: brushing the surface layer of the ECC specimens, using exposed aggregates on the ECC surface, and introducing grooves to the ECC specimens.

Microscopic skid resistance was evaluated by means of BPT following the ASTM E303-93 [1]. BPT is a dynamic pendulum impact-type test used to measure the energy loss when the rubber slider slides over the pavement surface. The test results are reported as British Pendulum Number (BPN). At least three specimens were evaluated for each mix, and a minimum of five tests was performed for each specimen. The tests were conducted in wet test conditions by spraying water on the specimens. BPT was also used to demonstrate the effects of macrotexture provided by the grooves. Previous studies mentioned that BPT is reliable for defining both microtexture and macrotexture [2, 3]. However, the macroscopic skid resistance was mainly evaluated by means of SPM, a volumetric-based spot test method with a sand patch. The method, which is based on ASTM E965-96, can be used to determine the (average) Mean Texture Depth (MTD) of pavement which reflects the macrotexture characteristic of the surface.

As can be seen from Table 7.1, various methods of surface treatments have been studied in this study. The effect of macrotexture on overall skid resistance is investigated by creating grooves on the PUTW surface. The grooves were manufactured with

Table 7.1 Investigation on skid resistance for surface treatments on laboratory

No	Objective improvement	Method	Precast	Cast in situ
1	Microtexture	Brushed ECC mixed with aggregates	Y	Y
2	Microtexture	Exposed aggregates on ECC surface	Y	Y
3	Macrotexture	Grooved ECC mixed with aggregates (precast with acrylic plate)	Y	
4a	Macrotexture	Grooved ECC mixed with aggregates (precast with Styrofoam)	Y	
4b	Macrotexture	Grooved ECC mixed with aggregates (precast with Styrofoam with larger gap)	Y	
5	Macrotexture and microtexture	Grooved ECC mixed with aggregates (precast with silicone rubber)	Y	
6	Macrotexture and microtexture	Grooved ECC mixed with aggregates (with concrete cutter machine)	Y	Y (requires heavy machinery)

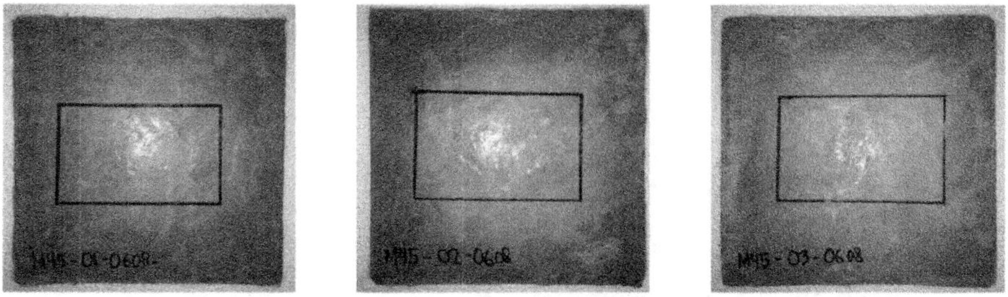

(a) (b) (c)

Fig. 7.1 Specimens of ECC M45

different experimental methods, including prefabricated and using grooves cutter. Table 7.1 summarizes methods used for precast and or cast in situ.

7.2.1 ECC Without Any Surface Treatment

A typical ECC M45 mixture design is adopted. This ECC design has the characteristics of a fly ash–cement (FA/C) ratio of 1.20 and a water-cementitious (W/C) ratio of 0.27. The mixing procedure started with cementitious materials (cement and fly ash) being mixed with sand and corundum. The amount of cementitious materials, sand and corundum, were kept constant. Water was added after those materials were mixed. Later, the HRWRA was added to the mixture until the desired characteristics were achieved, which was decided visually, both for the distribution of materials and the flow of the mixtures. Therefore, the HRWRA content was not kept constant. As the last step, the fiber (PVA) added to the mixtures. The whole mixing process was done in 8–10 min. All the mixtures were prepared using a small size standard mortar mixer with maximum mixtures of seven liters (Fig. 7.1). The mixture design of ECC M45, named as ECC-Cor 0 can be seen in Table 6.1.

7.2.2 Method 1: Brushed ECC Mixed with Aggregate

The objective of this method is mainly to improve the microtexture of ECC by mixing ECC with aggregates together. This process can be applied for both the "cast in situ" and "precast" method. Two aggregates are investigated: corundum and silica sand. ECC mixed with corundum is named as ECC-Cor, while ECC mixed with silica sand is named as ECC-SS. The mixture designs for both can be seen in Tables 6.1 and 6.2.

The ECC-Cor mixture was prepared in a planetary mixer. Cement, fly ash, and aggregates are mixed dry first. Water was then added, followed by the addition of

(a) (b)

Fig. 7.2 Specimens of **a** ECC-Cor, **b** ECC-SS

superplasticizer and mixed for 3–5 min until the mix achieved the required consistency. Lastly, the PVA fiber and aggregates were added and mixed for another 3–5 min. Specimens were removed from the mold after one day and were cured at normal laboratory air (25 ± 3 °C, 70%RH) for another 27 days before further testing. At the pre-determined age, either one or three days, the top surface of plate specimens was treated with a steel brush to expose the aggregates on the surface. The second mixture, ECC-SS, is designed with a different objective. It offers good skid resistance with a brighter surface layer for lesser impact to UHI and better visualization. Thus, the fine aggregate silica sand, which has a high hardness (slightly below corundum) and bright color, is used. The mixing process was done in a similar way to ECC-Cor (Fig. 7.2).

7.2.3 Method 2: Exposed Aggregates on ECC Surface

The second method, exposed aggregates on the ECC surface, is developed to maintain the original mechanical properties of ECC and to have good skid resistance by using fewer aggregates. Instead of mixing aggregates into the ECC mixture, the aggregates were distributed uniformly on the base of the plate molds, followed by casting the ECC (M45) on top of the aggregates. Plate specimens were removed from the mold after one day, followed by curing at normal laboratory air (25 ± 3 °C, 70%RH) for another 27 days before further testing (Fig. 7.3).

(a) (b)

Fig. 7.3 Specimens of exposed corundum aggregates on ECC surface

7.2.4 Method 3: Grooved ECC Mixed with Aggregates (Precast with Acrylic Mold)

The third method is the grooved ECC mixed with aggregates. The grooves made by using a precast by using a plate with space for the grooves. The plate is made of acrylic material with a thickness of 3.2 mm. The plate was made with two layers. The first layer is used as a base, while the second layer made for designed grooves. Both layers are glued to each other using glue for acrylic. Advantages of using acrylic are the material is relatively easy to handle, easy to cut with the laser cutting machine, and is light. The plate is later to be placed inside the mold (see Fig. 7.4). The ECC material is then to be poured afterwards.

However, the method was not successfully fabricating the grooves. The acrylic plate was stuck tightly on the grooves between the concrete. Due to this reason, the edges of each groove textures were broken when the acrylic demolded. Some acrylics were not able to be removed from the ECC samples. From the first experiment and its result, it can be concluded that acrylic should not be used to produce the grooves for ECC slabs as the material cannot be demolded easily. Therefore, BPT and SPM tests were not conducted on this method.

7.2.5 Method 4: Grooved ECC Mixed with Aggregates (Precast with Styrofoam Mold)

The experimental test for the fourth method was done with the help of Styrofoam material used as a mat. The Styrofoam was cut according to the required design with the help of an electronic hot wire foam cutting machine (see Fig. 7.5). The Styrofoam mat has a thickness of 3.2 mm and consists of two bonded layers. The first layer used as a base, and the second layer made for designed grooves. Styrofoam material was

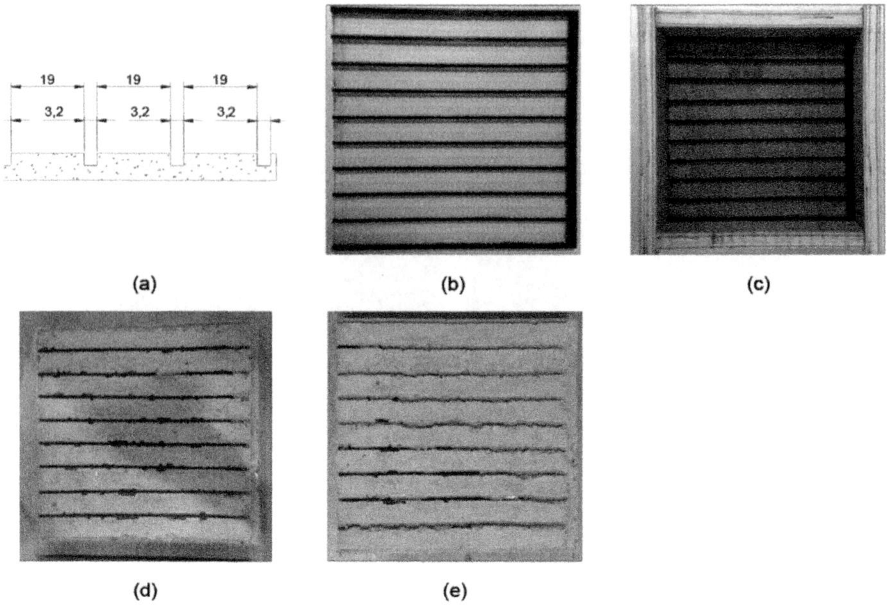

Fig. 7.4 Method 3rd: prefabricated grooves with acrylic plate **a** dimension of the design of grooves **b** acrylic grooving plate **c** acrylic grooving plate fits into mold **d** acrylic was stuck on the grooves of the ECC sample **e** broken edges of ECC sample

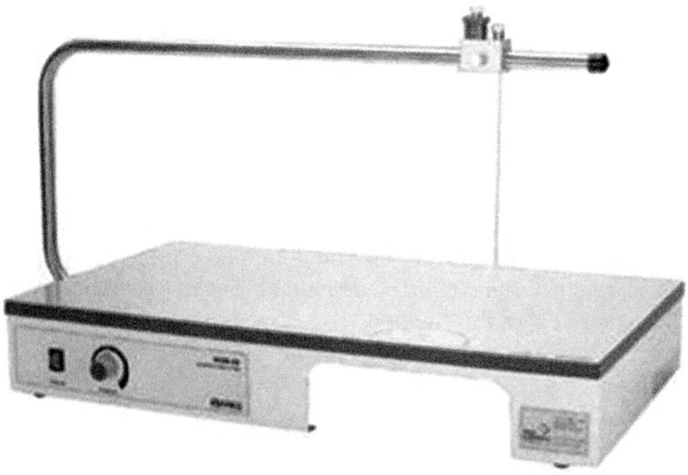

Fig. 7.5 Electronic hot wire foam cutting machine

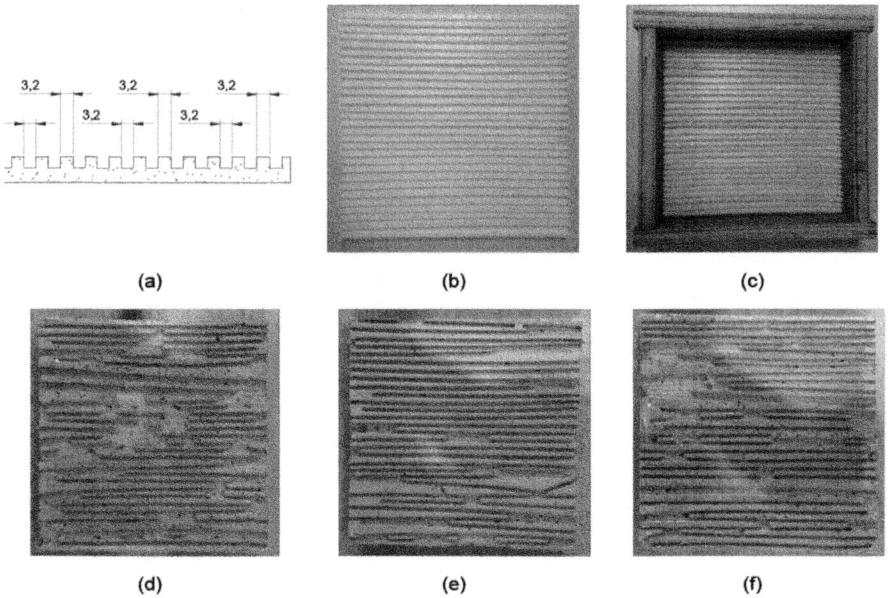

Fig. 7.6 Method 4th (1st design): prefabricated grooves with Styrofoam **a** design of grooves **b** styrofoam plate **c** styroforam plate fits into the mould **d–f** sample results

chosen as an alternative due to its flexibility, so it should be easily demolded. The cut Styrofoam then glued to another Styrofoam material resulting in a mat with tailored grooves.

Two different designs were prepared so they can be used as a comparison. The first design was made to imitate the grind-treated concrete pavement (small land area), which is illustrated in Fig. 7.6. As shown in Fig. 7.7, the second design was prepared to imitate the design of grooved pavement (larger land area).

As results, not all ECC samples were cast successfully. As can be seen in Fig. 7.6, the results from the first design, the grind-treated texture were not come up as expected. As the Styrofoam was too thin, it was moving all the time when the ECC was poured.

The second design with the fourth method has better results than the first design (grinding texture). The bigger land area between grooves makes it more stable, as the Styrofoam were less moving when the ECC was poured. Later, the Styrofoam can be easily removed by using Acetone. However, the results are not precisely done as some dimensions of the land area and the groove is different from the original plan.

In summary, using a Styrofoam material as a mat to produce grooves has a limitation when it comes to a smaller land area. As the ECC material is dense and heavy, it will easily move the Styrofoam. However, using Styrofoam for a wider land area results in a better sample. Yet, the grooves made of Styrofoam mat were not perfectly straight. The BPT and SPM tests are only conducted on the second design. Results are summarized in Table 7.3. The third method, with acrylics, shows no results as the

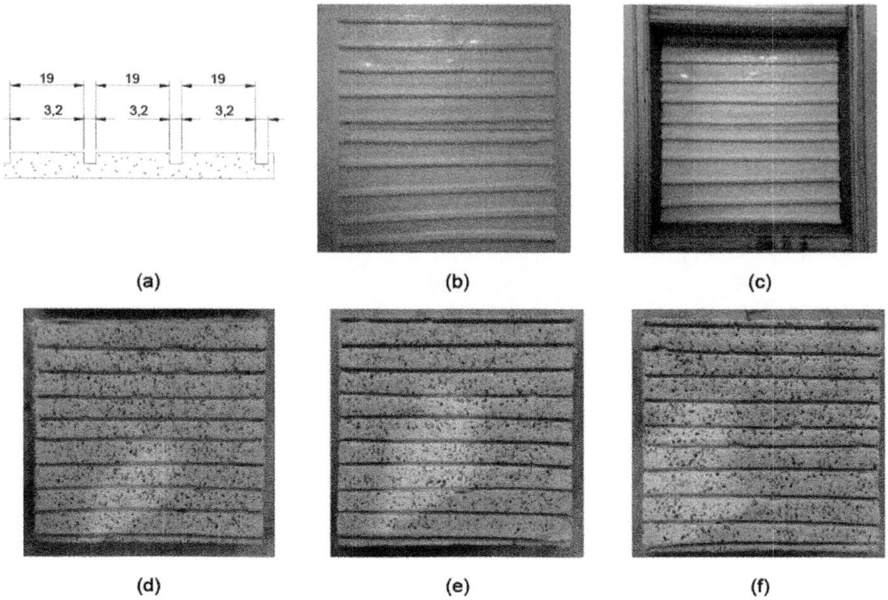

Fig. 7.7 Method 4th (2nd design): prefabricated grooves with Styrofoam **a** design of grooves **b** styrofoam plate **c** styroforam plate fits into the mould **d–f** results

specimens were not cast successfully. Thus, no tests were conducted. On the other hand, the fourth and fifth methods show promising results. Both Styrofoam and silicone rubber can be used to precast grooves for ECC specimens. However, it should be noted that precast with silicone rubber offers more advantages as it is easy to be made, more stable compared to Styrofoam. As results, BPN for longitudinal grooves is considered to remain the same, while BPN for transversal grooves increases. MTD for ECC with grooves is also increased about three times compared to the specimen without grooves.

7.2.6 Method 5: Grooved ECC Mixed with Aggregates (Precast with Silicone Rubber)

The experiment with the fifth method was done by preparing a silicone rubber mat. The first step is to prepare the wooden mold plate, which was cut with a laser cutting machine to produce precise holes as a negative mold for the silicone rubber. The thickness of the wooden plate is similar to the designed thickness of grooves, which is 3.2 mm. After the wooden plate is ready, silicone flowable liquid and curing agent (with a mix ratio of 100:2) is then mixed and prepared. The mixed silicone liquid then to be cast into mold with a wooden plate inside. The mixed silicone liquid will

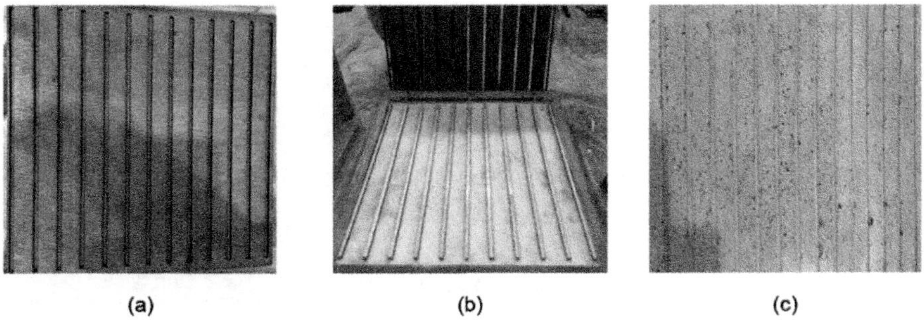

Fig. 7.8 Method 5th prefabricated grooves with silicone rubber: **a** negative wooden mold plate grooves with Styrofoam **b** casting silicone rubber for the prefabricated mold **c** results of grooved ECC sample by using mold made of silicone rubber

be hardened after 3–4 h at room temperature. The hardened silicone rubber is then placed into another mold, and it is ready to be used to cast the ECC (Fig. 7.8).

7.2.7 *Method 6: Grooved ECC Mixed with Aggregate (with Concrete Cutter Machine)*

The sixth method is the grooved ECC mixed with aggregate. The grooves were successfully implemented by using a concrete cutter machine. Parallel grooves were created on the surface of ECC-Cor 80 specimens (500 mm × 300 mm × 10 mm) by means of a concrete cutter machine. The concrete cutter was equipped with a blade of 350 mm in diameter and 3.2 mm in thickness. The blade position is adjustable, so it can slide forward and backwards, while the level can be fixed to make a groove with the specified depth. Four macrotextures were evaluated, and the corresponding groove dimensions are summarized in Table 37 (Fig. 7.9, Table 7.2).

7.3 Results

7.3.1 *Skid Resistance Performance of ECC and Modified ECC*

BPN of ECC M45 (ECC-Cor 0) and ECC-Cor # specimens with brushing and exposing aggregate methods are presented in Fig. 7.10. As can be seen, without the addition of corundum, ECC-Cor 0 has a BPN of 53 ± 9.1, which is lower than the required skid resistance of 55 for new roadways in accordance with the local

Fig. 7.9 Method 6th: Grooving with concrete cutter machine

Table 7.2 Groove dimensions for grooved ECC-Cor

Type	Land area (mm)	Groove (mm)	Height (mm)	Illustration
ECC-Cor 80-Gro 01	25.0 ± 0.5	3.2 ± 0.2	3.2 ± 0.3	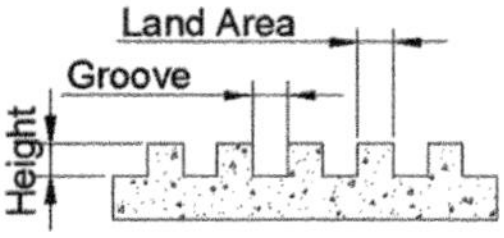
ECC-Cor 80-Gro 02	20.0 ± 0.5	3.2 ± 0.2	3.2 ± 0.3	
ECC-Cor 80-Gro 03	15.0 ± 0.5	3.2 ± 0.2	3.2 ± 0.3	
ECC-Cor 80-Gro 04	7.5 ± 0.5	3.2 ± 0.2	±0.3	

Table 7.3 Effect of grooves to microtexture and macrotexture with precast method

Specimen	Microtexture—longitudinal (BPN)	Microtexture—transversal (BPN)	Macrotexture, MTD (mm)
ECC-Cor 0	54.15 ± 2.50		0.09 ± 0.005
ECC-Cor 80	71.43 ± 2.02		0.36 ± 0.019
ECC-Cor 80-Gro (3rd method)	N/A	N/A	N/A
ECC-Cor 80-Gro (4th method)	67.93 ± 2.79	87.67 ± 4.78	1.03 ± 0.130
ECC-Cor 80-Gro (5th method)	72.00 ± 0.89	89.20 ± 0.75	0.86 ± 0.018

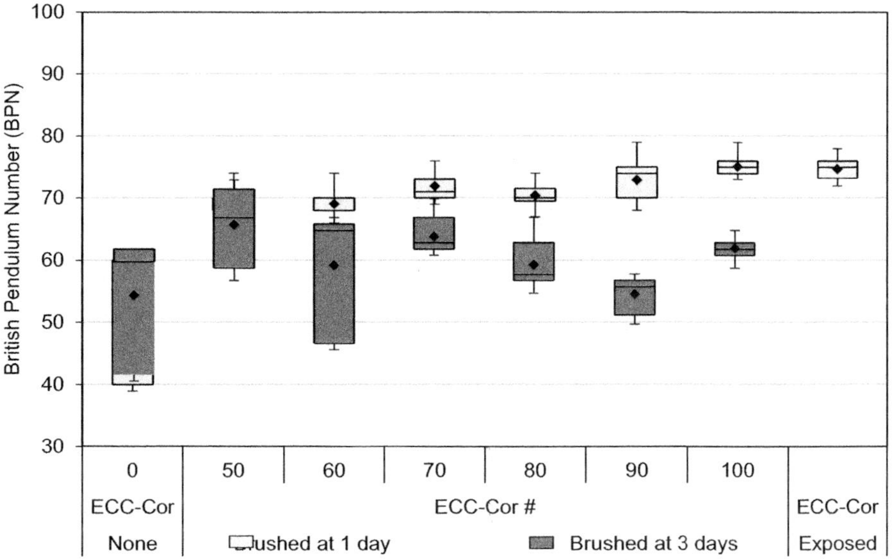

Fig. 7.10 BPN of ECC-Cor

standard set by the Land Transport Authority of Singapore. This suggests ECC-Cor 0 cannot be used for road application unless surface treatments are introduced.

ECC mixed with corundum (i.e., ECC-Cor 50 to 100) with surface exposure after three days shows some improvement in skid resistance as most of the results are above 55 BPN on average, except ECC-Cor 90. The moderate improvement may be attributed to the delayed surface exposure process at day 3, where the matrix, a fiber-reinforced paste, was very hard to be removed and thus, fewer corundum aggregates were exposed on the surface. It can be seen from the same figure that there is no clear correlation between the amount of corundum aggregates and skid resistance of ECC-Cor. This again may be attributed to the delayed surface exposure process at day 3, which is not able to expose the surface corundum evenly, as evidenced by the large variation of the BPN results.

A significant improvement of skid resistance of ECC-Cor with exposing process one day after casting was observed (BPN above 65). From Fig. 7.10, it can be observed that skid resistance (BPN) increases with an increasing amount of corundum used in the ECC-Cor mix. The ECC-Cor 100 possesses a BPN of 75, about 40% improvement compared to ECC-Cor 0. This is because the surface exposure process was carried out at day 1, and the paste can be removed easily and evenly, as evidenced by the reduced variation of the BPN results.

High skid resistance is also observed for the ECC-Cor specimens with the exposed method (BPN around 75, comparable to the ECC-Cor 100). Even with much less corundum was included in the ECC-Cor 0-S specimen (150 kg/m^3) as compared to other ECC-Cor mixes (e.g., 225 and 450 kg/m^3 for ECC-Cor 50 and 100 mixes, respectively). Since skid resistance is a surface property, ECC-Cor 0-S represents

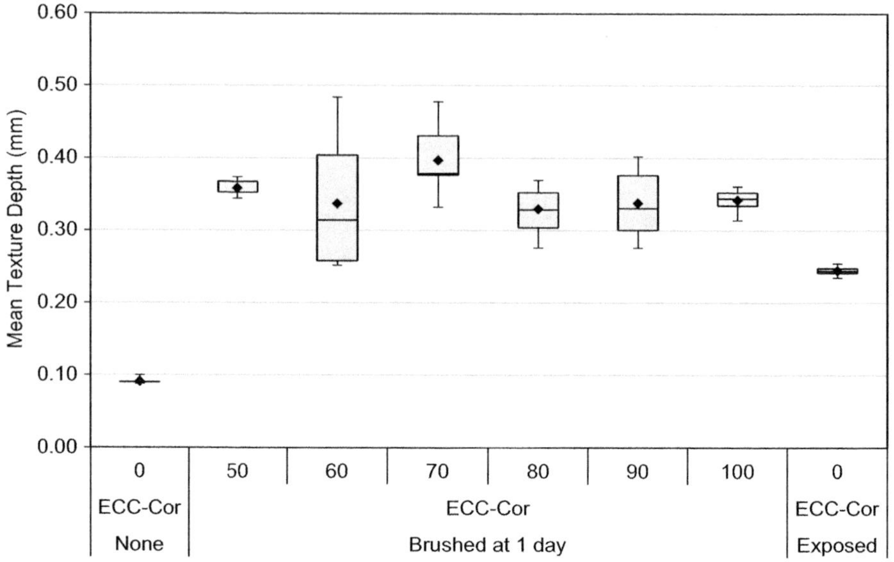

Fig. 7.11 MTD of ECC-Cor

a potential approach to strategically use corundum to improve the surface property while keeping the superior mechanical performance of the original ECC. Ultimately, this approach may be applied to both cast-in-situ and precast construction. BPN and MTD results can be seen in Figs. 7.10 and 7.11, respectively.

As shown in Fig. 7.11, the macrotexture of ECC-Cor 0 is very low, with a mean texture depth of only 0.09 ± 0.005 mm. Replacement of 50% corundum (ECC-Cor 50) remarkably increases the MTD to 0.36 ± 0.02 mm. Further increment of corundum dosage does not lead to continuous enhancement of MTD. On average, the MTD of ECC-Cor # is about 0.35 ± 0.05 mm, which is three times that of ECC-Cor 0.

BPN of ECC-SS specimens with the brushing method is presented in Fig. 7.12. As can be seen, without the addition of fine aggregate silica sand, ECC-SS 00 has a BPN of 54 ± 6.6 lower than the required skid resistance of 55 for new roadways in Singapore. This suggests ECC-SS 0 cannot be used for road application unless surface treatments are introduced.

A significant improvement of skid resistance of ECC-SS with exposing process one day after casting was observed (BPN above 65). The ECC-SS 01 possesses a BPN of 66, and ECC-SS 02 possesses a BPN of 79, about 20% and 45% improvement, respectively, compared to ECC-SS 00. However, more silica sand (ECC-SS 03) does not significantly improve the results compared to ECC-SS 02. As shown in Fig. 7.13, the macrotexture of ECC-SS 00 is very low, with the mean texture depth of only 0.09 ± 0.002 mm. An additional 10% of fine aggregate silica sand (ECC-SS 10) remarkably increases the MTD to 0.22 ± 0.04 mm. While an additional 20% of fine

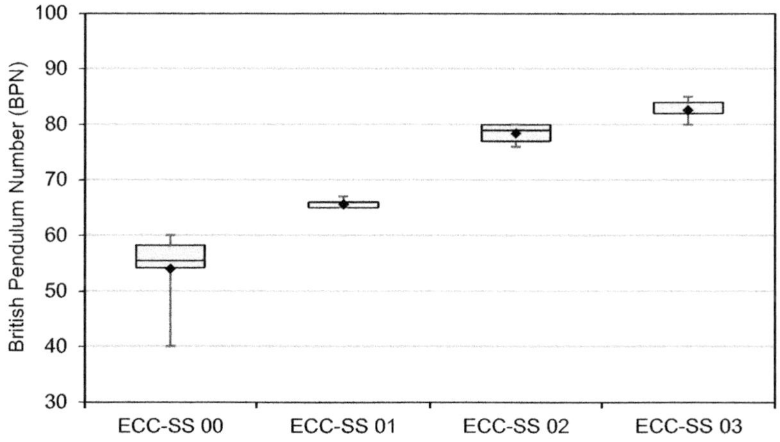

Fig. 7.12 BPN of ECC-SS

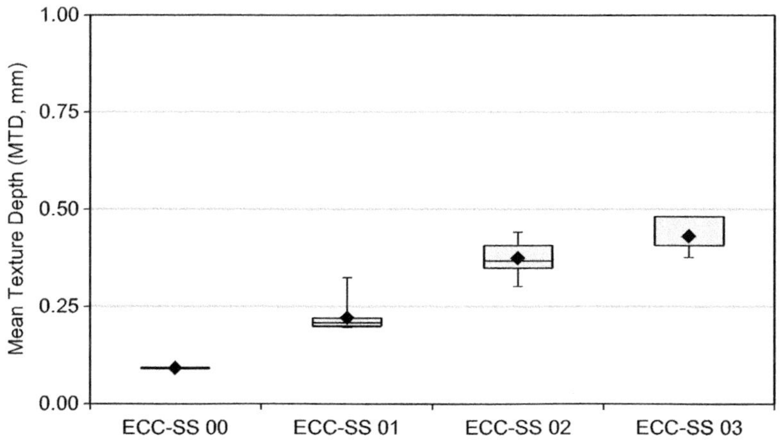

Fig. 7.13 MTD of ECC-SS

aggregate silica sand will lead the MTD to 0.37 ± 0.04 mm. Further increment of fine aggregate silica sand dosage does not lead to continuous enhancement of MTD.

During the laboratory tests, it was found that the fines are affecting the skid resistance. When the fines are covering the surface of the non-exposed specimens, the results of BPT on the wet condition are sometimes higher than the dry condition. But when the fines are washed away during the test on the wet condition, the results are decreasing.

In summary, the brushed ECC mixed with aggregates method comes with its advantages and disadvantages, including:

- It can fulfil the requirements of skid resistance, which means it enhances the safety factor of surface pavement being used by the motorist. Both mixture designs of ECC-Cor and ECC-SS can achieve high BPN (more than 55 BPN).
- It gives good skid resistance only when the brushing/exposing process done by one day. As longer than one day leads to hardened condition causes difficulties in removing fine material.
- It is trade-offs with mechanical behavior. As there is a certain volume of fine silica sand replaced by aggregates, the behavior of ECC is affected. Therefore, the amount of aggregates needs to be carefully designed.
- It is durable. The aggregates are well distributed in the matrix. Since the aggregates embedded inside the matrix of ECC, there will be another layer of aggregates once the aggregates on the surface are starting to wear off together with the mortar.
- It is a relatively easy procedure to cast ECC mixed with aggregates, followed by a brushing process.
- It means additional cost for adding aggregates into the mixture of ECC. As 50% replacement of fine silica sand, that covered around 225 kg per cubic meter of corundum.

The advantages and disadvantages of the exposed aggregates method include:

- It offers a solution for high skid resistance without affecting the mechanical behavior of ECC as it is a separate layer between the ECC the aggregates.
- It can fulfil the requirements of skid resistance, which means it enhances the safety factor of surface pavement being used by the motorist.
- It offers a lower cost compared to the previous method (mixing aggregates with ECC). The aggregate needed for this method is much less (150 kg/m^3) compared to the first (mixing aggregate with ECC) method, which requires at least 225 kg/m^3 (ECC-Cor 50) to 450 kg/m^3 (ECC-Cor 100).
- It could offer lesser texture durability compared to the mixing aggregate with ECC the method as the aggregates are only on the surface. Once the aggregates removed along with time/traffic, there will be no other aggregates that appear on the surface.

7.3.2 Skid Resistance Performance on Grooved ECC

Table 7.3 summarizes the BPN and MTF of ECC-Cor specimens with grooves with different precast methods. The third method, with acrylics, shows no results as the specimens were not cast successfully. Thus, no tests were conducted. On the other hand, the fourth and fifth methods show promising results. Both Styrofoam and silicone rubber can be used to create precast grooves for ECC specimens. However, it should be noted that precast with silicone rubber offers more advantages as it is easy to be made, more stable compared to Styrofoam, and it can be used repetitively. As results, BPN for longitudinal grooves is considered to remain the same, while BPN for transversal grooves increases. MTD for ECC with grooves is also increased about three times compared to the specimen without grooves.

Table 7.4 Effect of grooves to microtexture and macrotexture

Specimen	Microtexture—longitudinal (BPN)	Microtexture—transversal (BPN)	Macrotexture, MTD (mm)
ECC-Cor 0	54.15 ± 2.50		0.09 ± 0.005
ECC-Cor 80	71.43 ± 2.02		0.36 ± 0.019
ECC-Cor 80-Gro 1	71.40 ± 1.43	77.50 ± 1.57	0.74 ± 0.030
ECC-Cor 80-Gro 2	72.10 ± 0.83	88.60 ± 1.28	0.85 ± 0.030
ECC-Cor 80-Gro 3	72.60 ± 0.92	90.50 ± 1.28	1.08 ± 0.054
ECC-Cor 80-Gro 4	74.10 ± 0.70	92.80 ± 1.60	1.33 ± 0.059

Table 7.3 summarizes the BPN and MTD of ECC-Cor 80 specimens with different surface grooving treatment (i.e., ECC-Cor 80-Gro 1 to 4), ECC-Cor 0 and ECC-Cor 80 as the control. As can also be seen from Fig. 7.16, the average texture depth increased for the ECC mixed with corundum compared to the conventional ECC with MTD of 0.09 and 0.36, respectively. Macrotextures are much improved with additional grooves, ranging from 0.70 to 1.41 depending on the dimensions of the grooves. With a land area of 25.0 ± 0.5 mm, the MTD is 0.74 mm, which doubles the result for ECC-Cor. When the spacing (land area) of the grooves are shortened, the MTD of the macrotexture improved accordingly. For a land area of 20.0 ± 0.5, 15.0 ± 0.5, and 7.5 ± 0.5 mm, the results of MTD are 0.85, 1.08 and 1.33 respectively (Table 7.4).

The BPN of the grooved ECC-Cor 80 tested in the transversal direction (i.e., the grooves are perpendicular to the BPT swing direction) is higher than that of the ECC-Cor 80. Furthermore, the BPN increases with reducing the land area. ECC-Cor 80-Gov 4 with a land area of 7.5 ± 0.5 mm shows the highest BPN of 92.80 ± 1.60, representing a 71% and 30% enhancement compared to that of ECC-Cor 0 and ECC-Cor 80, respectively. This may be attributed to the increased macrotexture on the surface, which holds the movement of the rubber slider on the pendulum. The BPN of the grooved ECC-Cor 80 tested in the longitudinal direction (i.e., the grooves are parallel to the BPT swing direction); however, it remains similar to ECC-Cor 80 (Figs. 7.14 and 7.15).

The results of the SPM test are represented in Fig. 7.16. The test results are described with a box plot statistical method. There are six types of results. The "ECC" represents the results for conventional ECC. "ECC-Cor" represents specimens of ECC mixed with corundum. "Gro" represents the results for grooved specimens of ECC mixed with corundum, while the number after "Gro" indicates the groove dimensions, which are specified in Table 7.2.

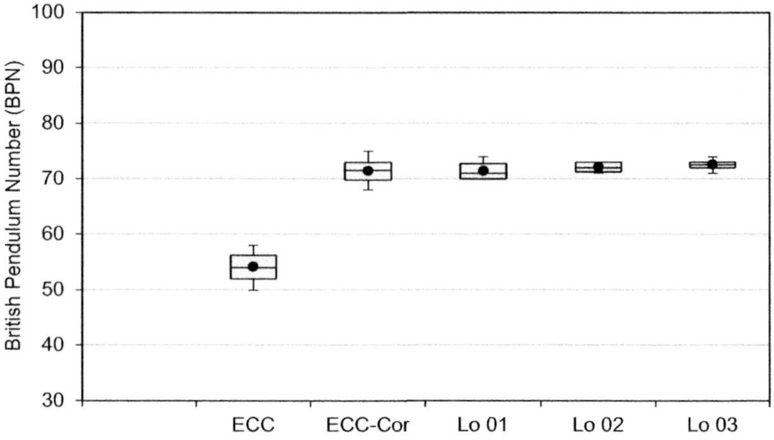

Fig. 7.14 BPN of ECC, ECC-Cor, and longitudinally grooved ECC-Cor specimens

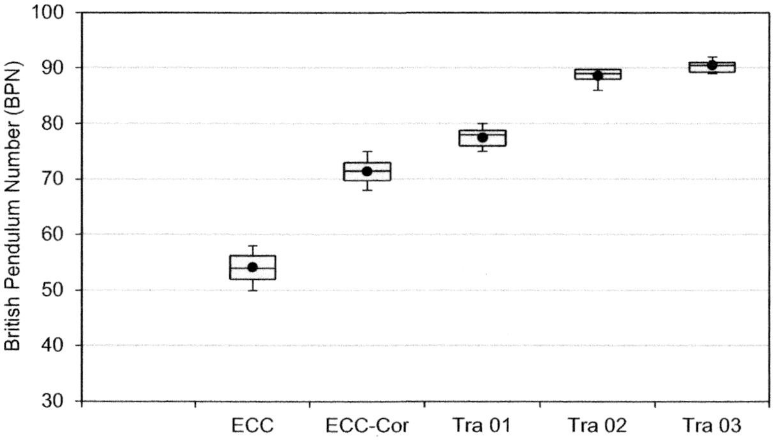

Fig. 7.15 BPN of ECC, ECC-Cor, and transversally grooved ECC-Cor specimens

In summary, the grooved ECC mixed with aggregates method comes with its advantages and disadvantages, including:

It can fulfil the requirements of skid resistance, which means it enhances the safety factor of surface pavement being used by the motorist. Both mixture designs of ECC-Cor and ECC-SS can achieve high BPN (more than 55 BPN).

It is durable. The aggregates are well distributed in the matrix. Since the aggregates embedded inside the matrix of ECC, there will be another layer of aggregates once the aggregates on the surface are starting to wear off together with the mortar.

It is a relatively easy procedure to cast ECC mixed with aggregates, followed by the grooving process. Both precast and cast in situ methods can make the grooves.

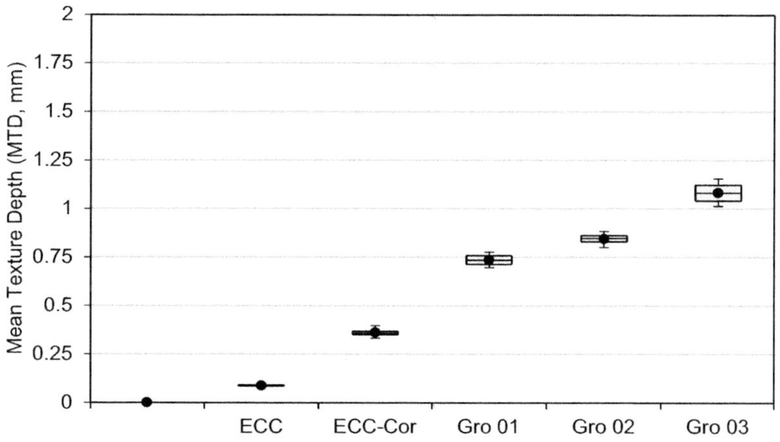

Fig. 7.16 MTD of ECC, modified ECC-Cor, and modified ECC-Cor with grooves

It means an additional cost for grooving pavement made of ECC.

References

1. ASTM E303–93 (1993) ASTM standard test method for measuring surface frictional properties using the British pendulum tester. ASTM International, West Conshohocken, PA
2. Karaca Z (2016) Reliability of BPT on macrotextured surfaces. Int J Innov Sci Res 5(1):611–616
3. Lee K (2004) Analyzing laboratory skid resistance test using finite element modeling. Doctoral Dissertation, National University of Singapore

Chapter 8
Study of Skid Resistance: Field Investigation

8.1 Introduction

This chapter presents the investigation on the full-scale application of modified ECC for pavement application. Data from various texturing methods applied in the field have been collected and then analyzed. The skid resistance tests conducted were including the British Pendulum Test (BPT) for the microtexture and the volumetric based Sand Patch Test (SPT) for the macrotexture. Different texturing methods applied to find the optimum solution to produce good surface properties for pavement application.

In comparison to the laboratory scale, there are more variables that need to be considered for a full-scale application of modified ECC, i.e., temperatures, mixing method, and finishing work. Test results show that the skid resistance of the surface pavement can be improved by adopting hard fine aggregates into the ECC mix, as it provides both the microtexture and macrotexture. It is also shown that texturing methods which was studied on laboratory scale tests can be further applied to full-scale application with consistent results.

8.2 Full-Scale Field Investigation Program

Microscopic skid resistance was evaluated by means of BPT following ASTM E303-93 [1], Standard Test Method for Measuring Surface Frictional Properties Using the British Pendulum Tester. BPT is a dynamic pendulum impact-type test used to measure the energy loss when the rubber slider slides over the pavement surface. The test results are reported as British Pendulum Number (BPN). BPT was also used to demonstrate the effects of macrotexture provided by the grooves. Previous studies mentioned that BPT is reliable for defining both microtexture and macrotexture [2, 3]. However, the macroscopic skid resistance was mainly evaluated by SPT, which is based on ASTM E965-96 to determine the Mean Texture Depth (MTD).

A. A. Bawono, *Engineered Cementitious Composites for Electrified Roadway in Megacities*, Springer Theses, https://doi.org/10.1007/978-3-030-88542-7_8

Table 8.1 Investigation on skid resistance for surface treatments on field application

No	Method	Precast	Cast in place
1	Brushed ECC mixed with aggregates (casting in place)		Y
2	Exposed aggregates on ECC surface	Y	Y
3	Precast modified ECC slabs (with wooden formwork)	Y	
4	Precast modified ECC slabs (with steel formwork)	Y	
5	Reverse precast modified ECC slabs	Y	

A full-scale application can be completely different from a laboratory test. Many ideal conditions which can be found in the laboratory is simply not existing in the field. The type of mixer, duration of mixing, temperature, way of casting, way of demolding are just some examples of factor that can simply affect the results. Various methods of surface treatments investigation in full-scale are summarized in Table 8.1.

8.2.1 *Method 1: Brushed ECC Mixed with Aggregates (Casting in Place)*

The first trial was done to check the workability of ECC material on a bigger scale as the main objective. The method chosen is casting in place, as this method considered the simplest one. Most of the area is done by normal ECC (M45), with only a quarter of the total area was treated with surface treatment by mixing aggregates, corundum together with the ECC mixture.

Two mixtures have been designed and prepared. A 1000 L of ECC was casting in situ on an area with a size of 6.0 m of length times 2.5 m of width. Another 80 L of ECC mixed with corundum was also mixed and then cast as a surface treatment on the smaller part. Two different mixers used. The 1000 L of ECC was mixed with the main mixer in the batching plant (Fig. 8.1a). The ECC with corundum was mixed with a smaller mixer with 100 L of capacity (Fig. 8.1b). The application of the modified ECC-Cor was made in two ways: mixed together with other components of ECC (illustrated in the green box) and exposed corundum aggregates on fresh ECC (illustrated in the yellow box). The finished surface was then levelled manually with trowels. The brushing process for the aggregates was done three different times. The first brushing process was done after four hours. The exposed aggregates method will be discussed in the next subchapter (Fig. 8.2).

Similar results to laboratory investigation, the corundum on the ECC surface is still covered with fibers and fine materials when they are mixed for the first time. Therefore, the corundum needs to be exposed. The exposing process can be done by using a brushing machine with a hard-plastic brush pad. The brushing machine specification can be seen in Table 8.2. The brushing machine and brushing plastic pad used can be seen in Fig. 8.3.

(a) (b)

Fig. 8.1 Concrete mixer **a** plant drum with 3000 L capacity, **b** with 100 L capacity

(a) (b)

Fig. 8.2 Finishing stage **a** surface levelling, **b** results with cast in situ

Table 8.2 Brushing machine specification

Voltage	Power	Speed	Weight	Disc diameter	Pad material
220 V/50 Hz	2 HP	175 RPM	55 kg	430 mm	Hard plastic

(a) (b)

Fig. 8.3 **a** Brushing machine, and **b** brushing pad

8.2.2 Method 2: Exposed Aggregates on ECC Surface

The second method, exposed aggregates, which is done by spreading aggregates onto the ECC surface, was investigated. The main objective of this method is to produce a structural component made of ECC material that has a good skid resistance but not bothering to change the prevailing mechanical properties of ECC. In this method, the mix design of ECC is used as a first layer (Fig. 8.5a). Aggregates are then spread onto the ECC surface to enhance the skid resistance (Fig. 8.5b). Less density of aggregates (kg/m^3) is used in this method compared to the previous method where aggregates mixed together with other ECC material. This method is applied to two applications: slab for pavement and footpath. The difference between mixing aggregates with ECC and spreading aggregates onto the surface of ECC can be seen in Fig. 8.4.

The first application, which is for the slab, was done with method cast in place. 1000 L of ECC was cast on an area with a size of 6.0 m of length times 2.5 m of width. The 1000 L of ECC was mixed with the main mixer in the batching plant (Fig. 8.1a). On top of ECC, corundum aggregates (100 kg/m^2) were spread.

This method was also applied to a footpath. The corundum aggregates (150 kg/m^2) were distributed on an area with 30 m length by 1.5 m width. The existing footpath was demolished and reconstructed with ECC and normal concrete with compressive strength of 45 MPa (used as a comparison). The concrete was treated with transverse tining. Reinforcement (M5) is used in some part of ECC and normal concrete, as this is the standard applied by LTA. In ECC, the first 15 m is reinforced, and the second 15 m is not reinforced. Meanwhile, the full length of the normal concrete part is equipped with reinforcement. Detailed design can be seen in Fig. 8.6.

The application is followed with skid resistance tests: BPT and SPM for both ECC-Cor and concrete material, located at five random spots. As can be seen from Fig. 8.18, both BPN and MTD for ECC-Cor with the spreading method is higher than conventional concrete due to corundum aggregates are well distributed on the surface.

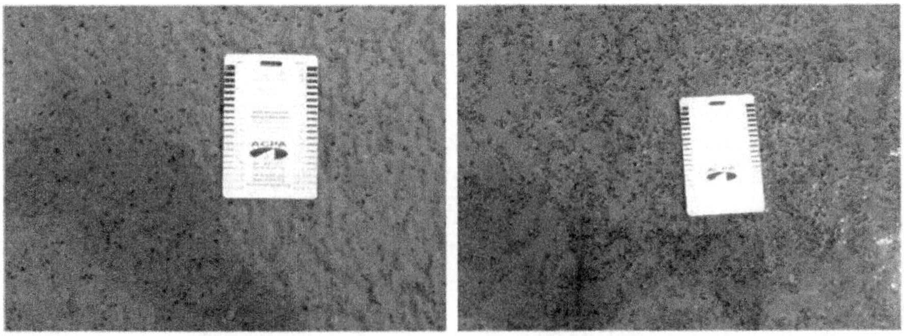

(a) (b)

Fig. 8.4 Texture of **a** of ECC-Cor, and **b** exposed corundum on ECC surface

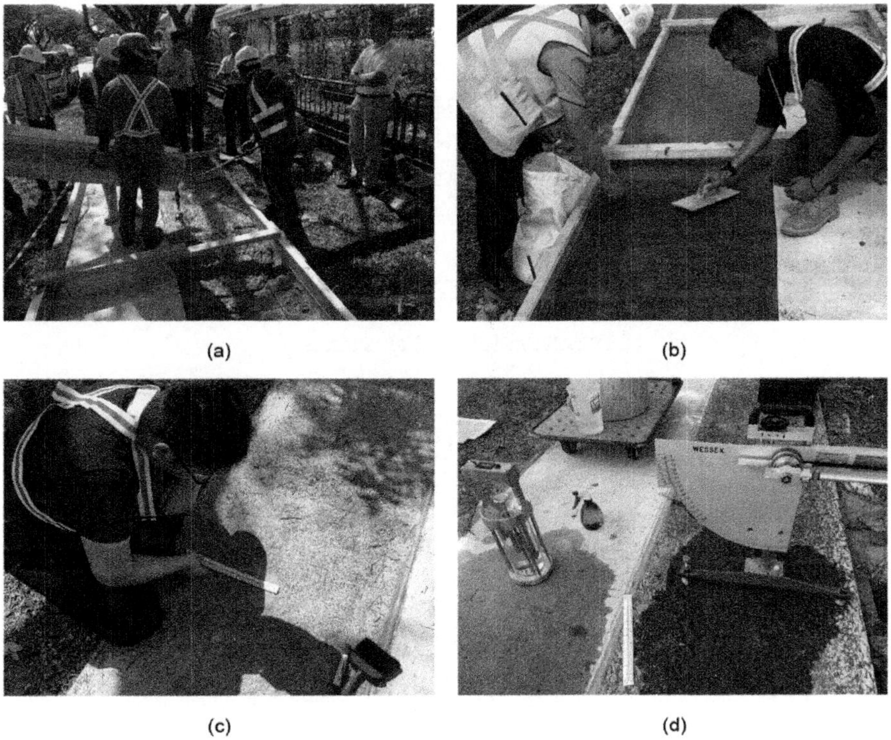

Fig. 8.5 Spreading aggregates onto ECC for footpath application **a** casting normal ECC, **b** spreading aggregates onto ECC surface, **c** SPM test, and **d** BPT test

8.2.3 Method 3: Precast Modified ECC Slab (First Trial with Wooden Formwork)

The objective of method 3 is to investigate the workability, the mixture design, and the results of the prefabricating method. Wooden formwork with the inner size of 3.6 m by 2.4 m has been prepared to fabricate the ECC slab (Fig. 8.7a). Two mixtures were prepared: the first mixture (normal ECC) is designed as a structural layer, while the second layer is modified ECC-Cor designed as a surface layer. A 1000 L of ECC was mixed in the batching plant and then cast into the wooden formwork. Another 80 L of modified ECC-Cor was also mixed and then cast as a surface layer (Fig. 8.7b). It is mixed with a smaller mixer with 100 L of capacity. The second layer must be done immediately with no more than 30 min after the first layer when the material is not hardened yet to produce better chemical bonding between layers. The finished surface of the slab was then levelled manually with a vibrator steel plate and trowels (see Fig. 8.7c). It is important to ensure that the surface to be flat. After one day, the surface layer of modified ECC-Cor was then brushed so the corundum will be

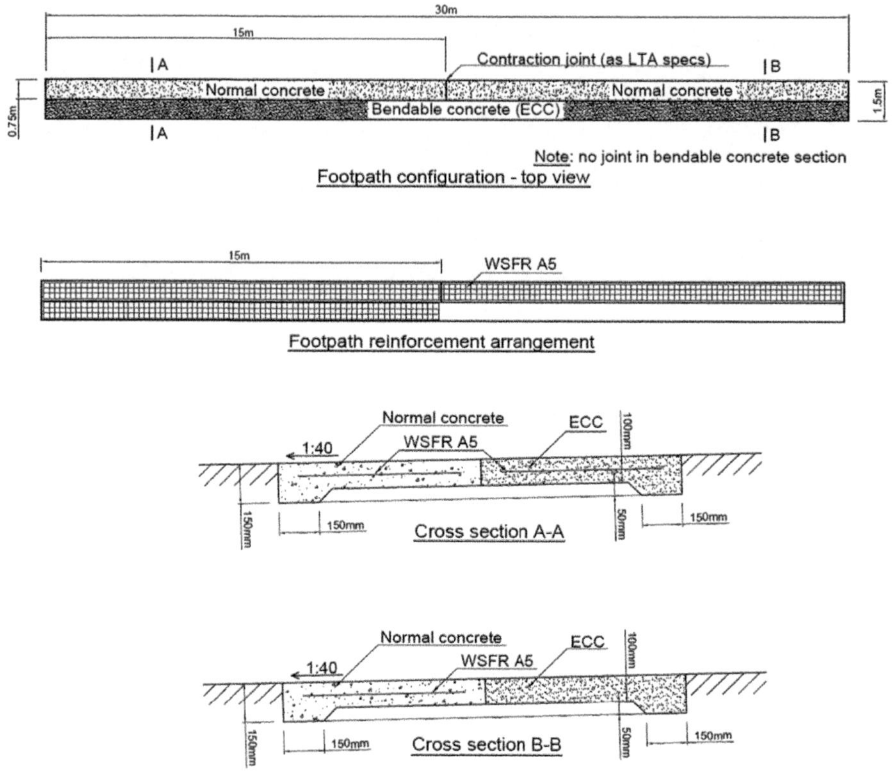

Fig. 8.6 Method 2: spreading aggregates onto ECC surface for footpath application

exposed. The exposing process was also done by using a brushing machine with a hard-plastic brush pad (see Fig. 8.19d).

There was a need to add a significant amount of plasticizer for the second batch due to the temperature difference between the night (the trial time) and the morning time (when the precast slab is prepared). According to weather.gov.sg, the minimum temperature in Tuas South area on 30th August was 25.1 °C, while the maximum temperature reached 31.3 °C.

8.2.4 *Method 4: Precast Modified ECC Slabs with Steel Formwork*

Method 4 is designed to improve the expecting results of previous method 3. Steel formwork is used to avoid the slab is becoming bent. Two steel molds were designed and ordered for the manufacturing of the bendable concrete slabs (Fig. 8.8a, b). Dimensions of the steel molds are 3.4 m × 2.4 m × 0.075 m and 1.7 m × 2.4 m

Fig. 8.7 Precast modified ECC mixed with aggregates **a** wooden mold 3.6 × 2.4 m^2, **b** pouring ECC into the wooden formwork, **c** levelling ECC-Cor surface, and **d** brushing process

× 0.075 m, respectively. The molds were designed such that the sidewalls can be removed right after the material harden to prevent the formation of shrinkage cracks. Silicone mattresses with circular holes (Fig. 8.8c) were also manufactured to be used in combination with the steel molds to create circular convex nodes on the bottom of the slabs, as shown in Fig. 8.8d. This is to increase the bonding of the interface between the bendable concrete slab layer and the underneath asphalt layer.

Seven large-scale mixings were conducted. The brushing process of the slab was done with the brushing machine. But the time of the brushing process was not always the same because of temperature. In the beginning, the brushing process was always done after one day. But then it found out for a bigger scale, the curing process needs more time, so the slab is hardened enough to be dry. When the slab is cast during a rainy day, more days are needed for the slab to be in dry condition. The mixing procedures were identical to that of previous large-scale mixing trials as introduced previously. As it has been discovered during previous trials, the water amount required to reach the desired workability may be affected by the weather of the day of casting. Table 8.3 summarizes the water to binder (w/b) ratio of each mix and the meteorological condition of each day of mixing. As can be seen, the w/b

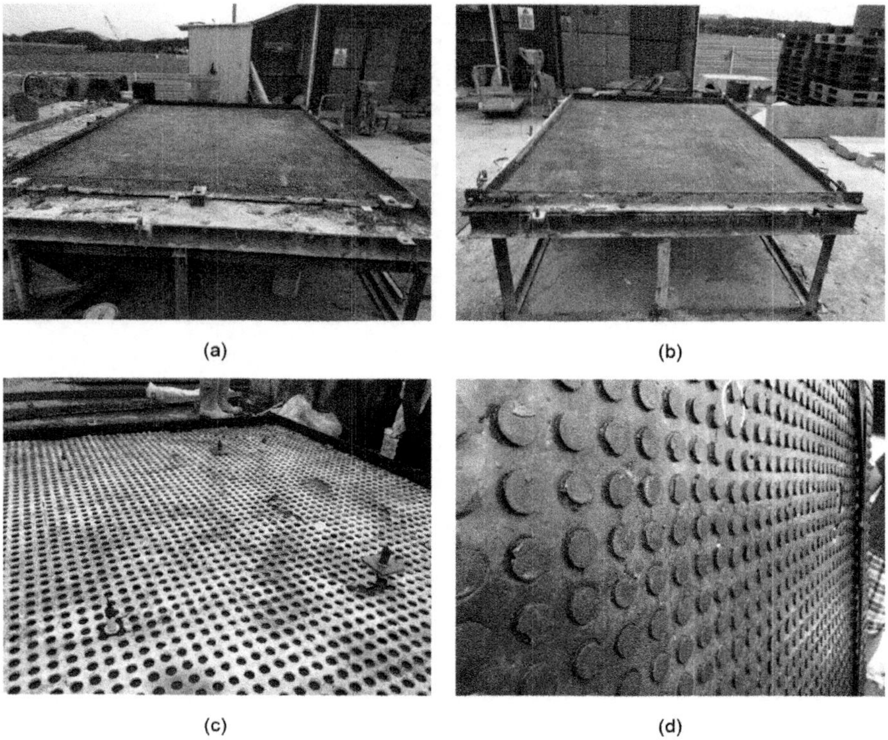

Fig. 8.8 Precast method **a** steel mold used for large slab, **b** steel mold used for small slab, **c** silicone mattress with circular holes and **d** circular convex nodes on the bottom surface of slab

Table 8.3 Mix designs and weather conditions

Date	Temperature (°C)	Rainfall on the previous day (mm)	Water/b ratio	Brushing (after × day)	Identification No. of slab (half size)	ID No. of slab (full size)
March 6, 2018	26.0–31.7, very dry	0.0	0.302	1	6	Failed
March 8, 2018	25.7–30.5, dry	5.4	0.269	1	4	11
March 13, 2018	26.1–33.3, dry	1.0	0.274	3	3	Back up
March 20, 2018	26.5–32.6, rainy	0.0	0.236	5	Back up	10
March 22, 2018	27.0–34.5, very dry	0.2	0.283	3	5	7
March 27, 2018	24.0–26.8, rainy	57.2	0.260	5	2	9
April 3, 2018	27.0–34, very dry	0.0	0.283	2	1	8

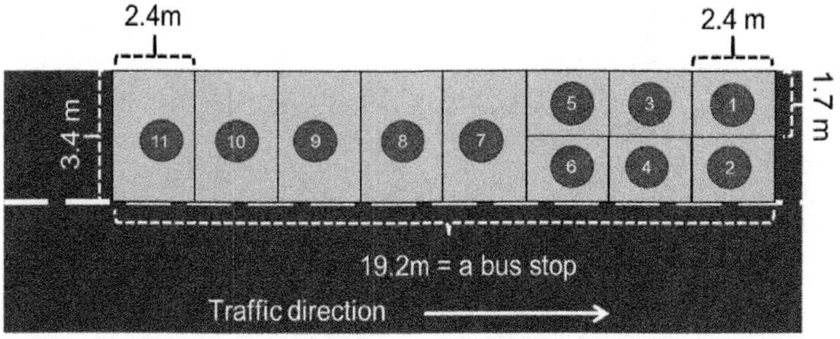

Fig. 8.9 Plan for modified ECC slabs on busbay application

ratio varies from 0.236 to 0.302 under different weather. The identification number of the slab can be seen in Fig. 8.9.

The process of producing and installation of a prefabricated modified ECC slab can be seen in Fig. 8.10. The first layer of ECC is cast into the formwork, then the second layer 10 mm ECC-Cor cast as a surface layer. After one day, the slab is demolded from the formwork, and then it is brushed to expose the aggregates. After the modified ECC slab is hardened, it can be deployed. Then, BPT and SPM tests were taken on site.

8.2.5 Method 5: Reverse Precast Modified ECC

The main purpose of the fifth method is to produce a better surface, which is flatter and smoother. By having a surface layer facing the formwork, it will create an ideal surface, which is found to produce the surface layer much better than the conventional precast method.

A half-size slab was made with the fifth method by using steel formwork with a dimension of 1.7 m × 2.4 m × 0.075 m. The molds were designed such that the sidewalls can be removed right after the material harden to prevent the formation of shrinkage cracks. Silicone mattresses with circular holes were also used to create circular convex nodes on the bottom of the slabs.

On the contrary to the previous method (normal prefabricate), the reverse method starts with the surface layer first. Apart from the reverse process differences, the lifting levelling device needs to be modified. The previous design of the lifting device (Fig. 8.11a) was mainly and can be only used for one direction, which is why the plate only embedded in the bottom area. The function of the plate is to hold the bear the weight of the slab when the slab is lifted. The plate is then working similarly when the slab needs to be levelled. However, the same design will be not working if reverse casting is proposed. For the reverse method, the lifting device should be able to work in both directions (going upwards and downwards). Two alternative designs can be used: using two plates embedded on both ends (Fig. 8.11b) or one plate embedded in the middle (Fig. 8.11c). As the surface layer on the bottom when

Fig. 8.10 **a** Casting first layer, **b** casting second layer ECC-Cor, **c** brushing process, **d** placing modified ECC slab, **e** SPM test, and **f** BPT test

the slab is cast, the plate of the lifting device needs to be on the bottom (surface layer) to bear the weight of the slab when it is lifted for the first time from the formworks. Then the slab needs to be rotated, so the surface layer will be on to facing upwards as there is another plate holding the slab from another surface, which is needed when the slab is then lifted and placed to the working site. This plate will also be needed for the levelling work.

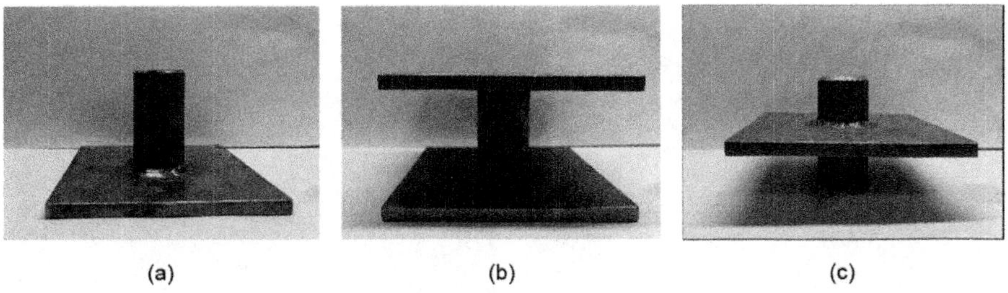

Fig. 8.11 Lifting levelling **a** design for precast, **b** and **c** design for reverse precast method

The modified lifting and levelling device, together with the pipe for grouting, are placed in the steel formwork (Fig. 8.12a). The modified ECC-Cor with a thickness of 10 mm is poured first onto the steel formwork. Then, the second layer, ECC material with a thickness of 60 mm, is poured on top of the ECC-Cor. The silicone rubber mat is then pressed onto the wet ECC (Fig. 8.12b). After the rubber mat is laid on the wet ECC, it is important to ensure that ECC material filled the holes, a vibrator machine can be used to help this process (Fig. 8.12c). The end results of the convex nodes can be seen in Fig. 8.12d. After the slab is hardened, the slab needs to be lifted and rotated. The surface needs to be brushed to expose the aggregates. End results of the surface layer can be seen in Fig. 8.12f.

8.3 Results

8.3.1 Skid Resistance of ECC in Large Scale

The results from casting in situ ECC on a large scale do not satisfy pavement application requirements. Not only the skid resistance is considerably low, but also it results in wrinkles, as shown in Fig. 8.13. The wrinkles due to the low workability of the ECC itself. The existence of fiber inside the mixture combined with a low water-cement ratio causing difficulties when levelling works need to be done on the field. Therefore, material ECC is not recommended for pavement application, except there is an improvement in the mixture design to improve the material's workability.

8.3.2 Skid Resistance of Brushed Modified ECC in Large Scale

Based on the workability, working with ECC mixed with corundum on the surface layer was much easier than normal ECC. Two reasons for the improvement of the

Fig. 8.12 Procedure of reverse prefabricating method **a** placing modified lifting levelling device, **b** silicone rubber mat place after material is poured, **c** vibrating silicone rubber mat, **d** end results of convex nodes, **e** lifting hardened slab **f** end results of surface layer for reverse method

workability are the higher density of the corundum material and the existence of the corundum as intermediate material between fine materials and fibers.

The brushing process for the aggregates was done three different times. The first brushing process was done after four hours. The work was relatively easy as the material was not completely hardened. However, as results, many corundum

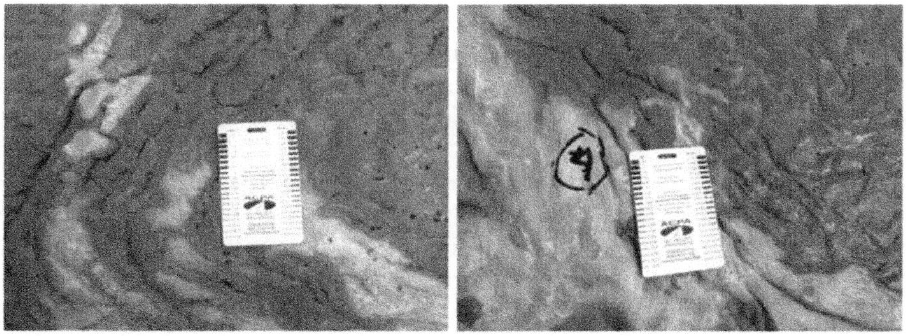

Fig. 8.13 Wrinkles on surface layer of normal ECC with cast in situ method

materials were found to be removed easily. The second brushing process was done five to six hours after the casting process. The fine materials can be removed easily, and corundum aggregates were exposed as a result. After six hours, the material is hardened enough. Therefore, the corundum materials were staying in their places. The last brushing process trial was done after 21 h. The fines still can be removed, but more efforts and times are required to get similar results as exposing process after six hours. The surface treatment with brushed ECC mixed with aggregates seems to work well. Time to expose between 6 and 12 h is suggested (Fig. 8.14).

Visual inspections and test measurements were conducted to check the performance of ECC with the cast in situ method. The visual inspections were done to check whether any cracks or deformation occur after some time. The performance of ECC was monitored after three days, seven days, 14 days, and 30 days. The second inspection was done after seven days. Similar condition as third days, neither cracks nor any deformation have been found after seven days. The third inspection was done after 14 days consists of visual inspection and test measurements on functional performance. Based on visual inspection, the structure performance of the pavement is in good condition as there is no crack, shrinkage, or deformation found on the site. Three tests on the functional performance have been performed: skid resistance related to the microtexture with British Pendulum tester, macrotexture with sand patch method test, and outflow with outflow meter.

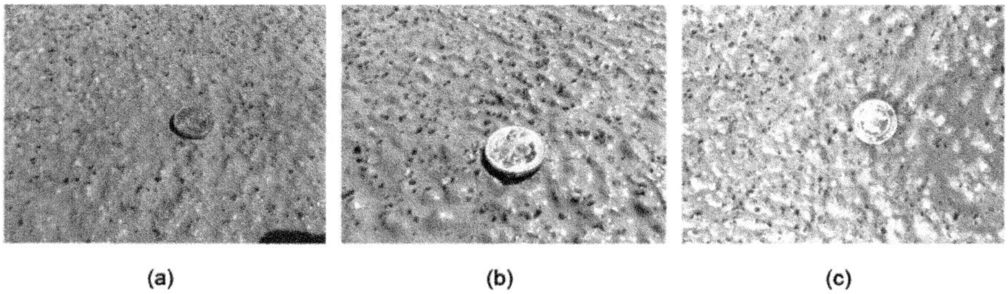

Fig. 8.14 Cast in situ ECC-Cor after **a** 4 h, **b** 6 h, and **c** 21 h

The fourth visual inspection was done after 30 days. A crack has been found in the middle of the slab, which is the location of the wheel trucks. The size of the crack is approximately 0.35 mm with a length of one meter. The crack occurred because there is no bond between the ECC layer and the layer below resulting in a hanging slab (Fig. 8.15).

Brushing ECC-Cor after 4 h causes many corundum were being removed easily. As results (shown in Fig. 8.16), the BPN is not improved significantly (average BPN is 56). Convincing results were acquired with exposing process after six hours. Fine materials were easily removed without removing the corundum as the matrix is sufficiently hardened. As consequences, BPN increased up to 80, with 72 as the average. The third trial of exposing process was done after 21 h resulting average BPN of 64 with 69 as the maximum BPN.

Based on the results, it is concluded that working on the surface after 21 h is still working but need more effort to reach similar results. A different method with

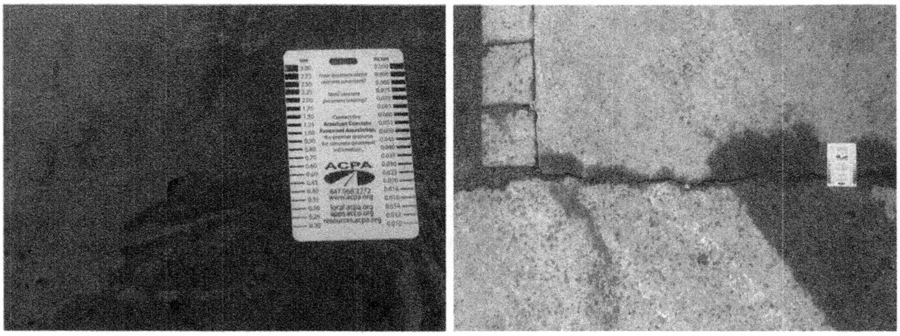

Fig. 8.15 Cracks on first trial cast in situ of ECC-Cor

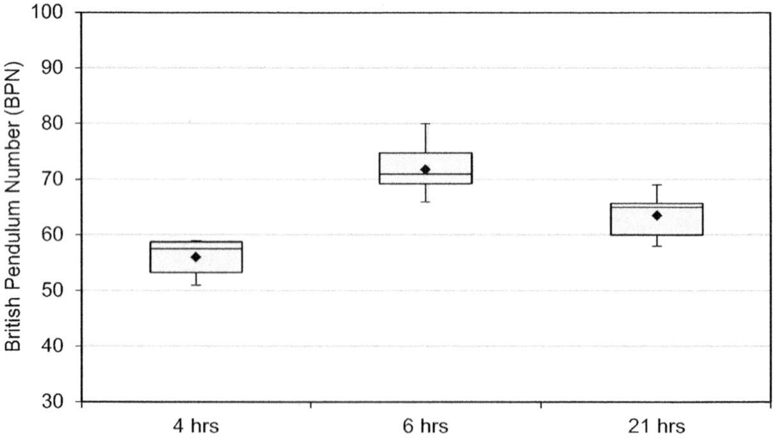

Fig. 8.16 BPN of brushed ECC mixed with aggreggates (method 1)

pouring corundum on top of ECC material increases the BPN to 60 as the average with 64 as the maximum. The microtexture results from field investigation are quite consistent with laboratory tests. 71.43 ± 2.02 BPN, while the average from field investigation is 72 BPN.

As can be seen from Fig. 8.17, results of the SPT of macrotexture on brushed ECC mixed with aggregates after four hours, six hours, and 21 h are 1.26 mm, 1.19 mm, and 1.22 mm, respectively. The results from the field test are more than double in comparison to the laboratory test, which is 0.36 ± 0.019 mm. The inconsistency occurs on a large scale, as the finishing of the ECC-Cor was not as smooth and flat as the smaller specimens. Therefore, the large-scale slab has higher macrotexture.

In summary, the brushed ECC mixed with aggregates method on a large scale comes with its advantages and disadvantages, including:

- It can fulfil the requirements of skid resistance, which means it enhances the safety factor of surface pavement being used by the motorist. Both mixture designs of ECC-Cor and ECC-SS can achieve high BPN (more than 55 BPN).
- It gives good skid resistance only when the brushing/exposing process done for 6–12 h. The latest of brushing process should be done within 24 h. As longer than one day leads to hardened condition causes difficulties in removing fine material. The faster process needs to be done on a larger scale because of ambient temperature.
- It is a trade-off with mechanical behavior. As there is a certain volume of fine silica sand replaced by aggregates, the behavior of ECC is affected. Therefore, the amount of aggregates needs to be carefully designed.
- It is durable. The aggregates are well distributed in the matrix. Since the aggregates embedded inside the matrix of ECC, there will be another layer of aggregates once the aggregates on the surface are starting to wear off together with the mortar.

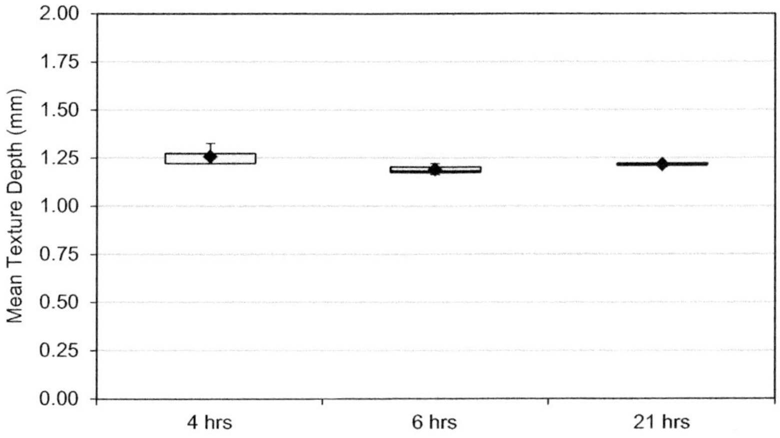

Fig. 8.17 MTD of brushed ECC mixed with aggregate methods (method 1)

- It is a relatively easy procedure to cast ECC mixed with aggregates, followed by a brushing process. However, a slight improvement on the mixture might result in a flatter and smoother surface.

8.3.3 Skid Resistance of Exposed ECC-Cor in Large Scale

The skid resistance performance of the exposed aggregates on the ECC surface can be seen in Fig. 8.18. The first trial was done for slab application. The second trial was done for footpath application, and it has a comparison with concrete M45 as both constructed in the same manner and time. The exposed aggregate on ECC for slab application results in lower BPN compared to the footpath as it is finished with fewer aggregates on the surface. On the contrary, the MTD for the slab is higher than the footpath due to the waviness of the un-flattened surface for the slab. In comparison to concrete with tining, the exposed aggregate results in higher BPN and MTD (mm).

In summary, the advantages and the disadvantages of the exposed aggregates on ECC on a large scale include:

- It offers a solution for high skid resistance without affecting the mechanical behavior of ECC as the corundum lays on the ECC layer. Further, it offers lower cost compared to the method of mixing aggregates with ECC. The aggregate needed for this method is much less (100–150 kg/m^3) in comparison to the first (mixing aggregate with ECC) method, which requires at least 225 kg/m^3 (ECC-Cor 50) to 450 kg/m^3 (ECC-Cor 100).
- It could offer lesser texture durability in comparison to the method of mixing ECC with aggregate as the aggregates only on the surface. Once the aggregates removed along with time/traffic, there will be no other aggregates that appear on the surface.
- It should be taken into account that both full-scale applications were made manually. This results in inconsistency for BPN, MTD, and visual/physical appearance. Proper machinery for distributing aggregates can be used to get more consistent results.

8.3.4 Skid Resistance of Precast Modified ECC-Cor Slabs in Large Scale

The first trial of the precast method for the modified ECC slabs on a large scale was done by using wooden formworks. BPT and SPM tests were conducted (Fig. 8.19e, f) on six random spots on the precast modified ECC-slab. As results, this application gives a microtexture of 71 ± 4.7 BPN and a macrotexture of 0.59 ± 0.08 mm, shown in Fig. 8.20.

The performance provided from this method has better results compared to cast in situ or spreading aggregates (Fig. 8.19a, b). However, the aggregates are not well

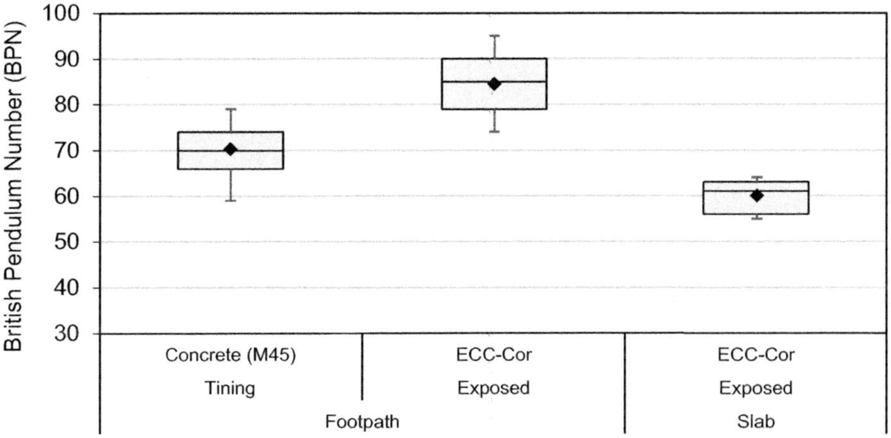

(a)

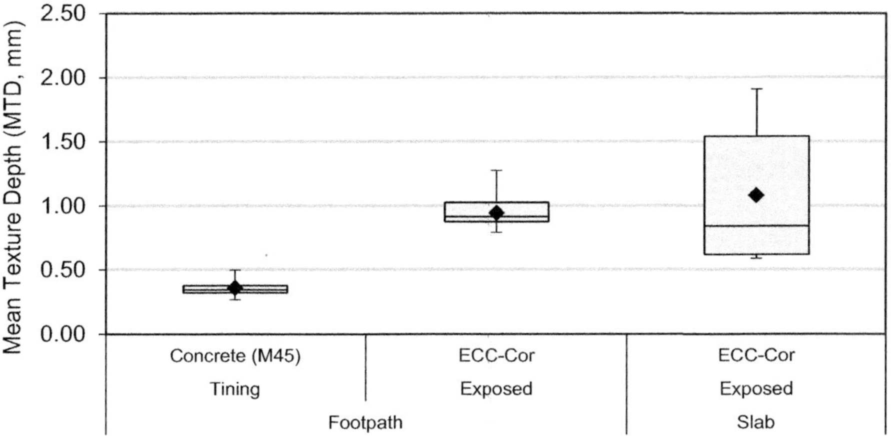

(b)

Fig. 8.18 Test results from exposed aggregates on ECC in large scale **a** microtexture in BPN **b** macrotexture in MTD, mm

distributed as the mixture is hard to work on. Some spots were identified to have minor roughness, as shown in Fig. 8.19c. Further, the slab was not ideal as the slab is bent because the wooden formwork was not able to bear the load of the slab itself. Not to mention that there are imperfect forms due to shrinking on the edge of the slab (Fig. 8.19d).

More precast modified ECC slabs were made by using steel formworks to avoid unnecessary deformation, which was occurred on slabs made with wooden formwork. About 12 slabs (details in Table 42) were cast. Five full-size and six half-size slabs were installed on busbay in Jurong Island. BPT and SPM tests were taken on site.

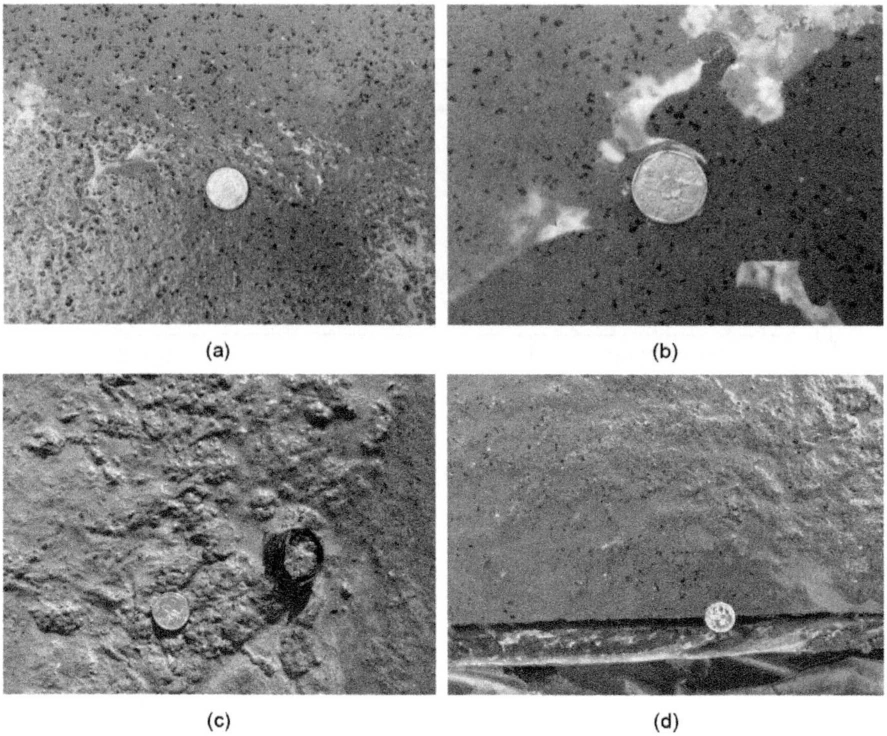

Fig. 8.19 Appearance of **a** ECC-Cor texture, **b** ECC-Cor texture (close look), **c** rough surface, and **d** gap between slab and wooden formwork

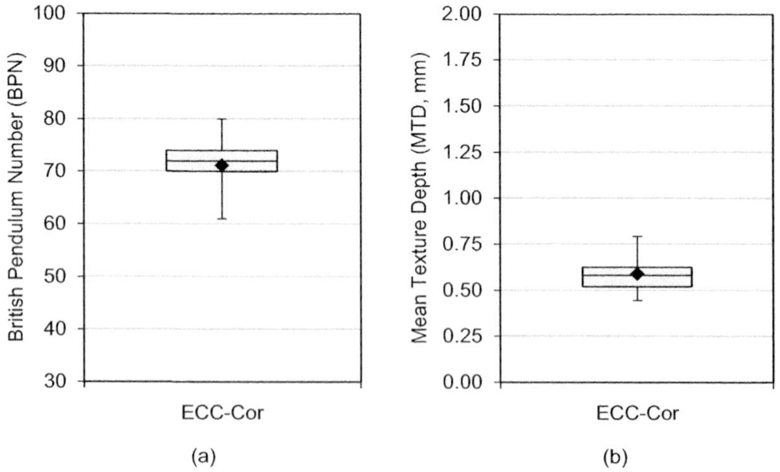

Fig. 8.20 Test results from precast modified ECC slab with wooden formwork **a** microtexture in BPN **b** macrotexture in MTD, mm

As can be seen in Fig. 8.21, various results are observed due to differences during mixing, casting, and finishing work. However, all slabs with the precast method result in more than 55 BPN, with an average of 68 ± 6.5 BPN for slab 02 to slab 11. The average for macrotexture is 0.44 ± 0.09 mm.

As has been mentioned, the reverse precast method is designed to solve previous issues faced, including deformation in the early stage, as well as inconsistency in a flat and smooth surface. As shown in Fig. 8.21 (slab 01), the reverse precast method

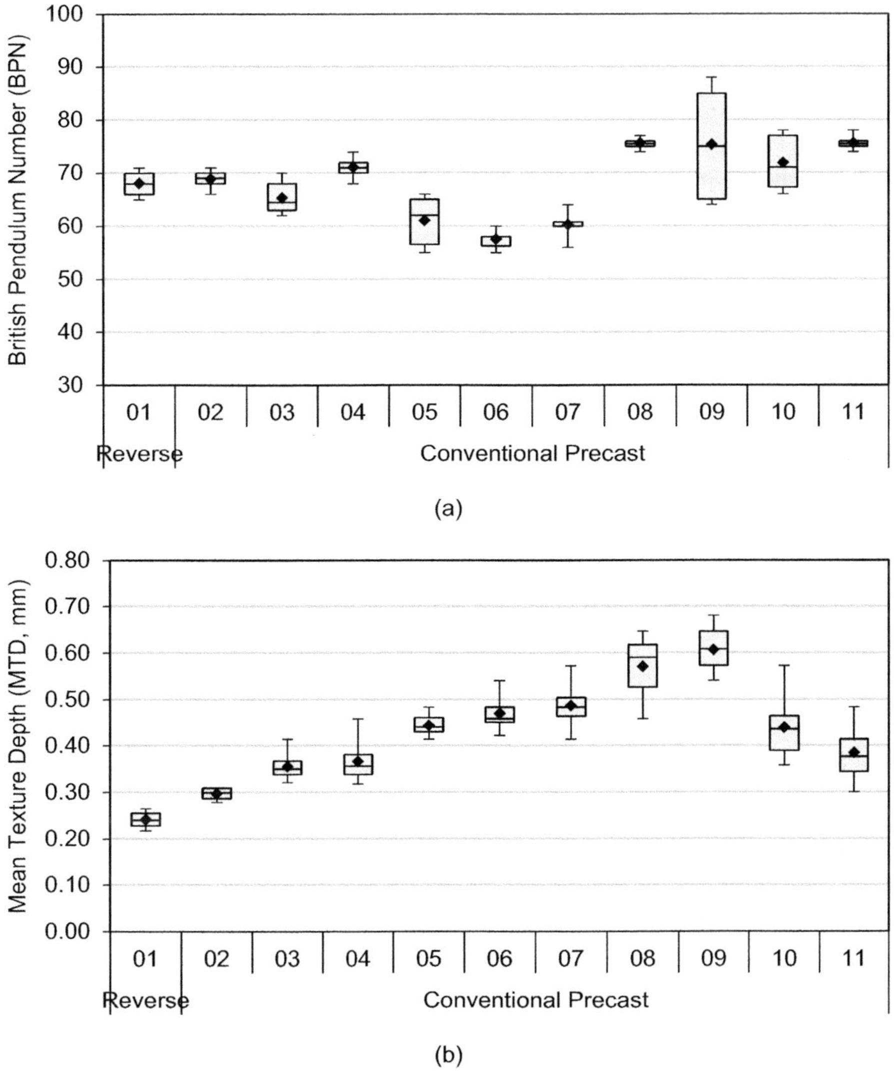

Fig. 8.21 ECC-Cor with precast and reverse precast method **a** microtexture in BPN, **b** macrotexture MTD in mm

for modified ECC gives results in microtexture 68 $\pm$ 2.45 BPN and macrotexture MTD 0.24 $\pm$ 0.02 as an average value. Further, this method improves the end result of physical appearance as the surface is flatter and smoother compared to other previous methods.

In summary, the precast ECC mixed with aggregates method gives conclusions including:

- It can fulfil the requirements of skid resistance, which means it enhances the safety factor of surface pavement being used by the motorist. This method gives high BPN (more than 55 BPN).
- Precast should be done with rigid-based material, such as steel. Wooden material allows deformation at an early stage.
- Precast with steel formwork gives good results and no deformation in an early stage. However, inconsistency is occurred due to several reasons: temperature during the mixing and casting process and the way of finishing works. The temperature issues can be easily solved if the mixing and casting process is done in indoors when the temperature is considered stable. As for the finishing work issue, several solutions can be sought, including improving the workability of the modified ECC by changing the mixture design, using standardized machinery to flatten the surface, or using the reverse precast method.
- The reverse precast method gives a flatter and smoother surface compared to conventional precast. Further, it provides the consistency for the slab manufacturing process. Consequently, the slab's manufacturing process needs to be adjusted, i.e., modification for the lifting and levelling design.

8.3.5 Comparison of "Cast in Situ" and "Precast" Method for Modified ECC

Cast in situ method for modified ECC slab has its advantages and disadvantages, including:

- It gives good skid resistance performance. However, the consistency might be low when manual work is done. Thus, machinery for finishing work is suggested.
- It offers a direct process. There are no requirements for formwork, handling and storage for the slabs, transport logistics. Consequently, it might less cost compared to the precast method.
- It does not need joints. As results, higher comfort and less traffic noise.
- It best for new construction as there is no consideration needed for curing time. However, for reconstruction work, this method might be less efficient compared to precast as it needs curing time, causing a longer time to open to traffic. Consequently, a higher life cycle cost is incurred.

Precast method for modified ECC slab has its advantages and disadvantages, including:

- It gives good skid resistance performance with good consistency.
- It offers an indirect process. There are many requirements for formwork, handling and storage for the slabs, transport logistics. Consequently, it might boost the cost compared to the cast in situ method.
- It is equipped with joints. In results, lesser comfort and higher traffic noise.
- It best for roadway reconstruction as there is a high consideration for the timing to be open to traffic. Consequently, lower life cycle cost is incurred.

References

1. ASTM E303-93 (1993) ASTM standard test method for measuring surface frictional properties using the British pendulum tester. ASTM International, West Conshohocken, PA
2. Karaca Z (2016) Reliability of BPT on macrotextured surfaces. Int J Innov Sci Res 5(1):611–616
3. Lee K (2004) Analyzing laboratory skid resistance test using finite element modeling. Doctoral dissertation, National University of Singapore

Chapter 9
Study of Skid Resistance: Numerical Modelling 3D FEM

This chapter describes FEM model development on analyzing and evaluating texturing methods to improve the skid resistance for pavement made of ECC material. Finite Element Method (FEM) model can be used to solve mathematical problems by translating partial differential equation into algebraic equations by representing the real structure into subdivision or parts called finite elements.

ANSYS as finite element software offers several computational algorithms for dynamics analysis. The development of the model is using ANSYS Version R17.2 Academic License. The objective is to develop a 3D FEM model according to the BPT test to measure the skid resistance of ECC material for PUTW. The BPT mechanism is based on a dynamic movement to measure the friction of the sample. Therefore, a dynamic analysis is needed. ANSYS offers various analyses such as Explicit Dynamics, Transient Structural, and Rigid Dynamics.

Explicit dynamics analysis is used to determine the dynamic response of a structure due to stress wave propagation, impact, or rapidly changing time-dependent loads. Momentum exchange between moving bodies and inertial effects are usually important parameters. Furthermore, this analysis can also be used to model mechanical movements that are highly nonlinear. Explicit dynamics analysis can be used to model both rigid and flexible bodies. However, this type of analysis cannot be used to model complicated mechanisms and joints, which are offered by transient structural analysis and rigid dynamics analysis. Explicit dynamics analysis is more efficient to study model on time scales of less than one second. For a longer time of duration events, the transient structural analysis should be considered.

The transient structural analysis, also called time-history analysis, can be performed for rigid and flexible bodies. This type of analysis is used to determine the dynamic response of a structure under the action of any general time-dependent loads. It can be used to determine the time-varying displacements, strains, stresses, and forces. The time scale of the loading is such that the inertia or damping effects are important. This analysis can be used either for linear or nonlinear. Rigid dynamics analysis can be used to determine the dynamic response of an assembly of rigid bodies, which is linked by joints, springs, and other body interactions. This analysis

A. A. Bawono, *Engineered Cementitious Composites for Electrified Roadway in Megacities*, Springer Theses, https://doi.org/10.1007/978-3-030-88542-7_9

Table 9.1 Inputs and outputs for 3D FEM model of BPT

Input parameters	Output variables
Material properties and element dimension	Sliding distance
Spring damping and stiffness	Time
Friction	Position (probe)
Body contact and interaction	BPN
Mesh design	

is suitable to study the kinematics of pendulum movement. However, rigid dynamic analysis has the limitation that both inputs and outputs are only related to forces, moments, displacements, velocities, and accelerations. While all parts are rigid that there are no stresses and strain results produced. In this research, rigid dynamics analysis is used.

Inputs are needed for the 3D FEM Model of the BPT simulation. Those inputs include material properties, spring damping, friction, body contact and body interaction, and mesh design. Data for inputs are collected from samples as well as references from another research. Through a significant number of simulations, results from the BPT model are collected and analyzed: sliding distance, time, and position. Those data will be analyzed further to get BPN as the result. Inputs and outputs from the 3D FEM model are summarized below (Table 9.1).

9.1 Material Properties

Material properties are determined in the early stage of ANSYS development. Materials used for the BPT test include aluminum for the pendulum, rubber for the rubber plates, and ECC for the sample tested, are used as inputs in the Engineering Data module.

Aluminum material is used for the pendulum of the BPT. It has properties of elastic modulus of 72.400 MPa, Poisson's ratio of 0.33, and density of 2700 kg/m^3.

The rubber slider is specified according to [1]. The rubber was modelled as a hyperelastic material. Though rubber is categorized as a viscoelastic material with a non-linear stress–strain relationship, during the test, it is reasonable to assume that the rubber remained wholly elastic with one compression modulus due to small deformation [3]. A recommended rubber modulus is from 3.22 to 3.29 MPa. In this model, the rubber modulus value of 3.29 MPa was used. A Poisson's ratio (μ) value of 0.47 for rubber slider is used, which was recommended [7]. In this model, the density of the rubber used is 1100 kg/m^3, which is based on laboratory measurement [4].

ECC has an Elastic Modulus of 18–34 GPa with Poisson's ratio of a range from 0.15 to 0.20 [5, 8]. In this 3D FEM model, the value of 22 GPa and 0.15 were adopted for elastic modulus and Poisson's ratio, respectively. The densities of ECC samples were calculated: 2040 and 2190 kg/m^3 for the ECC mixed corundum with Pour and

Table 9.2 Material properties for BPT model

Material	Aluminum	Rubber	ECC
Elastic modulus (MPa)	72.400	3.29	22.000
Poisson ration	0.33	0.47	0.15
Density (kg/m^3)	2700	1100	2100

Precast Method. Therefore, for this model, an average value of 2100 kg/m^3 has been used for the density of the specimen. This is complying with typical Portland cement mortar, which normally has a density of around 2000–2200 kg/m^3 (Table 9.2).

9.2 Geometric Model

As has been mentioned in the preliminary design, a simplified three-dimensional model of the British Pendulum Tester and a sample of ECC surface were created with the DesignModeler tool in ANSYS [2]. The dimension of every part of the BPT model is set accordingly to the actual BPT tester. The 3D FEM model of BPT is shown in Fig. 9.1.

The pendulum arm and the supporting arm are made of aluminum, which are designated as a beam. There are a lever arm and a lever system connecting the pendulum arm and the slider. The lever arm and the lever system also made with aluminum material. While the lever arm and the lever system connected by the spring. The spring behavior can be determined by its stiffness and damping coefficient, which

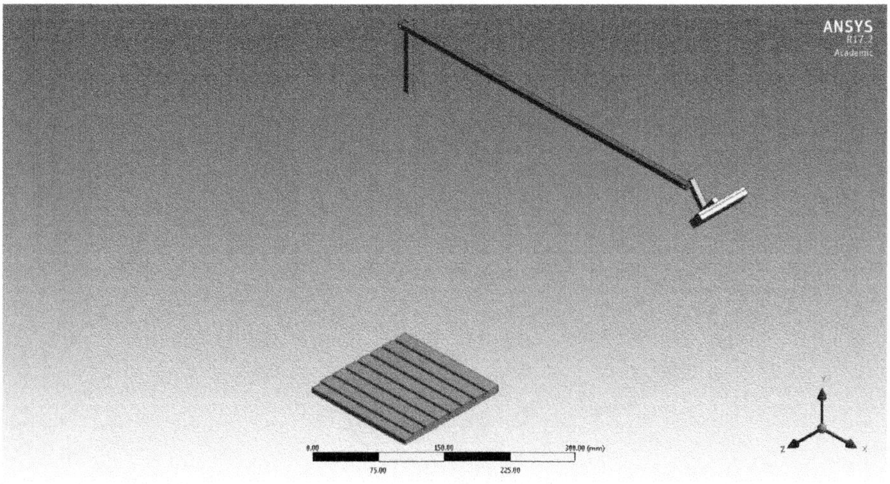

Fig. 9.1 3D FEM model of British Pendulum Tester

have to be adjusted. The stiffness and the damping were adjusted through verification of laboratory measurement and results obtained from the finite element simulation.

The cap of the rubber slider is also made of aluminum. The rubber slider was modelled with rubber with the approximate dimension of 76 mm by 25 mm by 6 mm (length × width × thickness). The total mass is 1.486 kg.

The ECC specimen model has the dimension of 150 mm × 150 mm × 10 mm. As specified in [1], the tester shall be capable of providing a slider contact path of 125 ± 1.6 mm for tests on flat surfaces. Therefore, the position of the sample can be, and it has to be adjusted to achieve a sufficient contact length between the slider and the sample.

9.3 Mesh Design

Mesh design in finite element analysis plays an important role on results. According to ANSYS, bad quality mesh can cause convergence difficulties, bad physic description, and diffuse solution. Therefore, a mesh design for FEM model should be carefully designed. The main objective of this 3D FEM model is to simulate the behavior of the friction on the surface sample. Therefore, mesh design is specifically designed only for the sample surface. The meshing was created with the help of Body Sizing and Face meshing tools to create a rectangular shape with maximum element size of 2 mm. This dimension was chosen considering the size of Maximum Aggregate Size in ECC mixture design. A dense mesh is selected for the rubber slider, especially at areas along the edge where contact was created between the rubber slider and the ECC. The meshing comprised 57,370 nodes and 18,312 elements. The meshing for smooth and grooved samples can be seen in Fig. 9.2 and Fig. 9.3, respectively. While meshing for rubber slider is shown in Fig. 9.4.

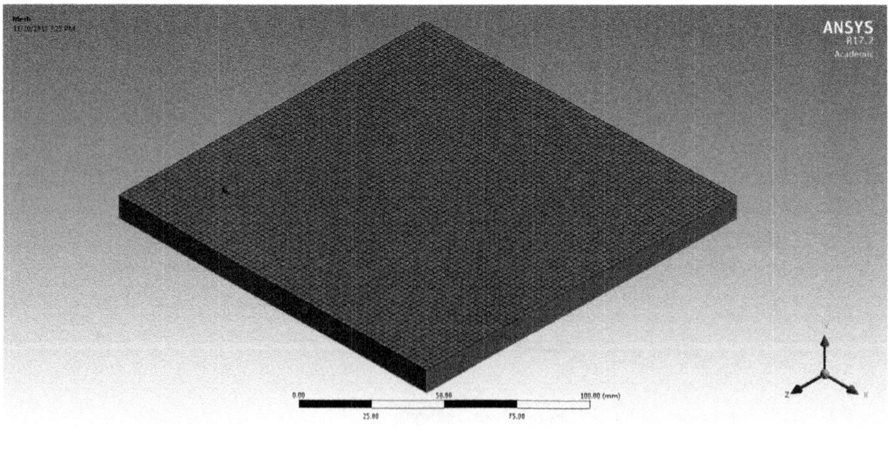

Fig. 9.2 Mesh design for smooth (ECC) sample

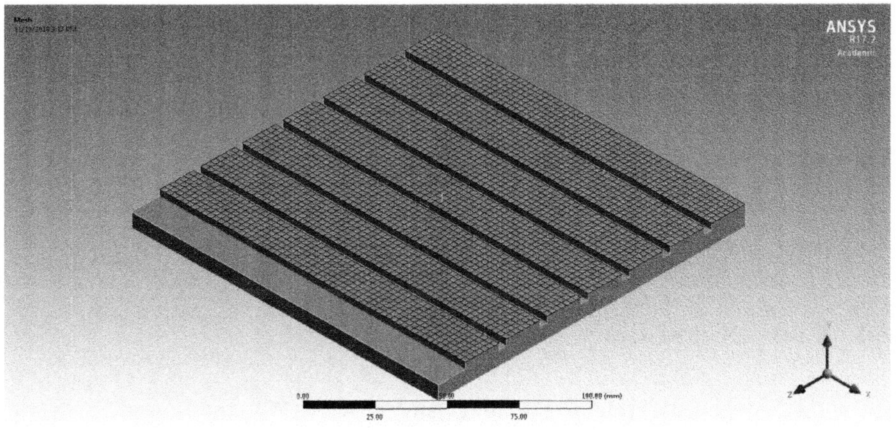

Fig. 9.3 Mesh design for grooved (ECC) sample

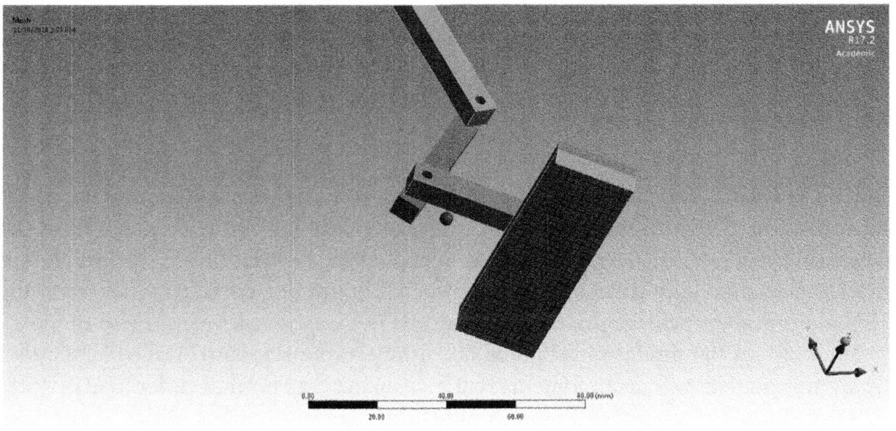

Fig. 9.4 Mesh design for rubber slider

The quality of the mesh has been checked in the ANSYS program. ANSYS recommend keeping minimum orthogonal quality above 0.1. Another parameter is the aspect ratio, which can be calculated by the length divided by the height. This model has an aspect ratio of 1.0024 as average. Based on the results from these parameters, the model shows good quality. The results for meshing are summarized in Table 9.3.

Table 9.3 Quality of meshing for 3D FEM model of BPT

Criteria	Value	Standard criteria
Element quality	0.99895 ± 0.0043	0–1 0 is worst, 1 is best
Orthogonal quality	0.99996 ± 0.0012	0–1 1 indicates a perfect cube or square; while 0 indicates element has a zero or negative volume
Aspect ratio	1.00308 ± 0.0353	The best possible quadrilateral aspect ratio is 1

9.4 Contact and Body Interaction

The slider contact path between the rubber slider and (ECC) sample surface is simulated with a setting type of manual contact with type of Force Frictional Sliding. The ECC sample surface is set as contact bodies, while the rubber slider is set as target bodies. The target surface is intended as a rigid surface, while the contact surface is a deformable surface.

9.5 Model Validation

The 3D FEM model of BPT has been validated with few parameters, including: (1) the validation of the mass of the model, (2) the validation of the distance from the center of gravity of the pendulum to the center of the oscillation, (3) the validation of sliding contact path distance, and (4) the calibration of correlation between the BPN measurement with real BPT tester and BPN measurement with FEM model.

The mass of the model is validated according to the standard ASTM E303, that the spring and the lever arrangement shall give an average normal slider load of 1500 $\pm$ 100 g. In this model, the pendulum, the slider, and the mount slider, has a total weight of 1500 g.

The validation of the distance from the center of gravity of the pendulum to the center of the oscillation shall be 411 $\pm$ 5 mm, which is in accordance with the standard [1]. While on this developed 3D FEM Model, the distance from the center of gravity of the pendulum to the rotation center is 411.2 mm.

The validation of the sliding contact path distance between the rubber slider and the sample surface is set according to the required standard of the BPT standard, which is 125 $\pm$ 1.6 mm. The verified result is shown in Fig. 9.5. In a real situation, this condition can be achieved by adjusting the position of the slider with the vertical height control knobs. On this 3D FEM model, parameters are considered to achieve this condition: the position (height) of the slider, the angle of the rubber slider and the spring coefficient for the rubber slider (Figs. 9.5 and 9.6).

The calibration of the correlation between the BPN measurement with the real BPT tester and BPN measurement with the FEM model was done in this study. A method was proposed by comparing the energy loss between laboratory measurement and

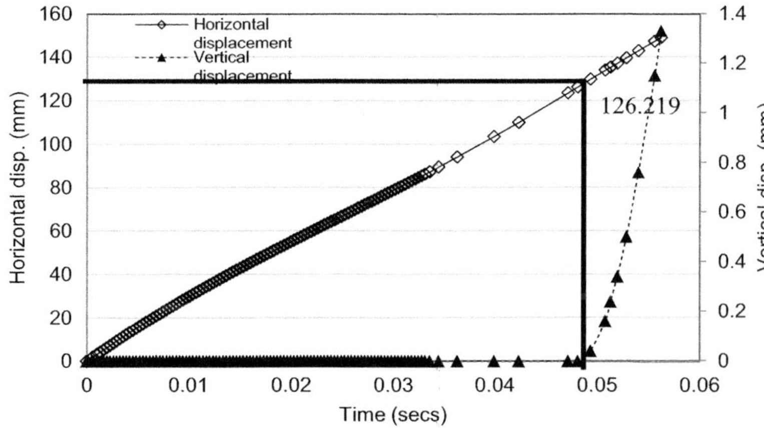

Fig. 9.5 Verification of sliding contact path distance for FEM model

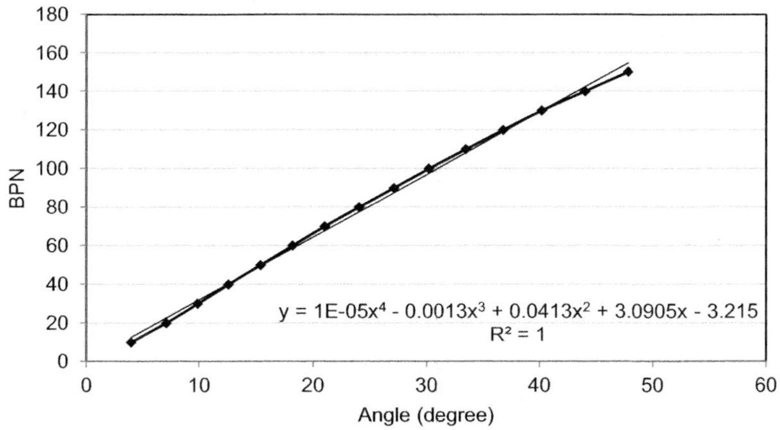

Fig. 9.6 Correlation between degree of angle of the pendulum and BPN

finite element analysis [6]. However, slight differences ~10% for results are expected. On this FEM model, the analysis is done by using the correlation between the degrees of the angle of the pendulum of BPT with the BPN values. Therefore, the BPN is calculated based on the position where the pendulum stops during the simulation. The correlation is derived to an equation below:

$$PN = 0.00001 xdeg^4 - 0.0013 xdeg^3 + 0.0413 xdeg^2 + 3.0905 xdeg - 3.215 \quad (9.1)$$

where: BPN_r = British Pendulum Number
deg = angle.

9.6 Results

The angle of the pendulum is calculated by using the probe tools in ANSYS. Afterwards, the BPN can be determined with the proposed equation. The results are summarized in Table 9.4.

The FEM model was validated by comparing the BPN results from the FEM analysis with BPN results from laboratory tests. The sample used for validation is ECC-Cor 80. Samples were prepared for both flat and grooved texture. This mixture design was chosen due to their skid resistance was one of the highest compared to another specimen. The comparison is summarized in Table 9.5. From the results, it can be seen the difference between the FEM model and laboratory test is below 5%, a considerably good model.

The 3D FEM model was conducted to predict the BPN of ECC samples. Validation and calibration of the model were performed with consideration of key parameters of the BPT mechanism: the mass, the center of gravity of the model and the sliding distance of the rubber slider. The simulation results show a good correlation with laboratory measurements.

The development 3D FEM model of BPT was done by using a Rigid Dynamic analysis. This analysis is used to determine the dynamic response of an assembly of rigid bodies linked by joints and springs, which are also the main parts of the BPT. However, Rigid Dynamics analysis has some limitation consist of: (1) inputs and outputs are limited to forces, moments, displacements, velocity and accelerations,

Table 9.4 Results of BPN from the finite element analysis

Mixture	Condition	Texture	FEA results		
			Position	Angle	BPN
ECC-Cor	Dry	No texture	−405.56	29	97
		Longitudinal grooving	−411.9	28	92
	Wet	No texture	−432.83	21	70
		Longitudinal grooving	−435.59	20	67

Table 9.5 BPN results comparison between experimental measurement and FEM model

Mixture	Test condition	Texture	BPN results		Percentage difference (%)
			Laboratory measurement	Finite element analysis	
ECC-Cor	Dry	No texture	97	97	0
	Dry	Longitudinal grooving	96	92	4.17
	Wet	No texture	70	70	0
	Wet	Longitudinal grooving	68	67	1.47

(2) all parts are rigid, that the system cannot analyze the stresses and the strain. It can only analyze forces, moments, displacements, velocities and accelerations, (3) viscous damping can be taken into account through springs, (4) another advantage is the solver is tuned to automatically adjust the time step.

References

1. ASTM E303-93 (1993) ASTM standard test method for measuring surface frictional properties using the British pendulum tester. ASTM International, West Conshohocken, PA
2. Bawono AA, NguyenDinh N, Lechner B, Yang E-H (2019) Study of finite element model on skid resistance for pavement made of engineered cementitious composite. Structural engineering, mechanics and computation, Cape Town, South Africa, September. CRC Press, London
3. Hills DA, Nowell D (1993) Mechanics of elastic contacts. Butterworth-Heinemann, Oxford
4. Lee K (2004) Analyzing laboratory skid resistance test using finite element modeling. Doctoral Dissertation, National University of Singapore
5. Li VC (2003) On engineered cementitious composites (ECC): a review of the material and its applications. Adv Concr Technol 1(3):215–230
6. Liu Y, Fwa TF, Choo YS (2003) Finite-element modeling of skid resistance test. J Transp Eng 129(3):316–321. https://doi.org/10.1061/(ASCE)0733-947X(2003)
7. Moore I, Purushothaman N, Heaton B (1988) Three-dimensional elastic finite element study of the skid resistance of grooved pavement. Int J Numer Meth Eng 26:437–452
8. Seracino R (2004) FRP composites in civil engineering. CICE 2004, Adelaide, Australia, 8–10 December. A.A. Balkema, London

Chapter 10
Study of Surface Water Drainage: Laboratory Test

This chapter presents the laboratory test on the surface water drainage performance of pavement made of modified ECC material. The mixture design of modified ECC incorporating hard aggregates is developed. A standard test with an outflow meter was utilized to investigate the surface water drainage performance. Subsequently, a tailored experimental was set to investigate the water depth/water film thickness on the surface of the pavement [1].

Surface water drainage is another important consideration in pavement design. The surface water drainage performance of pavement is divided into two mechanisms (PUB [8]: water surface runoff (water flowing above the surface pavement) and water infiltration (water infiltrates through the pavement). Water pond on the pavement surface produces excess water, which leads to hydroplaning effect and endangers the driver. Further, any water on the surface pavement might interrupt the traffic, reduce the grip, and limit the visibility due to splash and spray [2], which are amplified on an area with high rainfall intensity [5], Werner and Peck 2010). For further consideration, Singapore is a tropical city that receives approximately 2400 mm of precipitation annually and up to 19 high rainfall intensity days a month (PUB [8]. In comparison, data provided by the NOAA National Climatic Data Center, which is collected by weather stations throughout each state from 1971 to 2000, specified that Texas, North Carolina, and Louisiana experience average annual precipitation of roughly 734, 1279, and 996 mm, respectively [3]. The inclusion of coarse aggregates in concrete mix design introduces macrotexture significantly reduces the risk of hydroplaning effects on tire-pavement [7]. Besides surface texture, pavement geometry, i.e., cross-slopes and grades, can also affect surface water drainage performance [10]. Previous studies reported the correlation between water outflow and (macro)-texture, and few studies investigated the effect of pavement geometry on the surface water drainage. Furthermore, the effects of surface texture and pavement geometry on water drainage performance of ECC pavement are still not clear, and no model has been reported.

Singapore does not have a standard related to the measurement of surface water drainage. However, several standards by the Department of Transportation (DOT) in the USA specify a requirement for outflow times which can be used as a baseline.

A. A. Bawono, *Engineered Cementitious Composites for Electrified Roadway in Megacities*, Springer Theses, https://doi.org/10.1007/978-3-030-88542-7_10

Neither standard on water film thickness (WFT) can be found in Singapore. However, according to the Austroad [9], the requirement for water film thickness is:

- maximum water film depth of 2.5 mm (desirable) to 4.0 mm (absolute) for:
  - o Sections where design speed ≥80 km/hr,
  - o Approaches to and exits from intersections and roundabouts,
  - o Intersections, roundabouts, and steep downhill sections,
  - o Merge and diverge sections for entry ramps/overtaking lanes/climbing lanes,
  - o Superelevated curves (particularly those approaching limit curve speed).
- Maximum water film depth of 5 mm (desirable and absolute) for all other situations.

The test results of skid resistance show that the surface of pavement made of ECC can be improved by adopting hard fine aggregates into the ECC mixture, as it provides both the microtexture and macrotexture. However, there is still a risk of hydroplaning. Surface texturing methods can be introduced to reduce the risk of hydroplaning by improving the drainage performance as well as the skid resistance of ECC pavement. Therefore, studies on ECC's functional performance on roadway application needs to be extended, particularly on the surface water drainage performance since high rainfall intensity may lead to hydroplaning [2, 7]. Finally, an improvement on the surface of ECC is proposed to enhance the surface water drainage performance.

10.1 Laboratory Test Program

10.1.1 Materials and Mixture Design

Type I Portland cement, class F fly ash (FA), micro-silica sand (mean and maximum grain size of 110 μm and 200 μm, respectively), polyvinyl alcohol (PVA) fibers (length of 12 mm, a diameter of 39 μm, a nominal tensile strength of 1600 MPa), a high-range water-reducing (HRWR) were used to prepare the mix. Corundum with a maximum particle size of 2.38 mm was used as fine aggregates for improving the surface textures of ECC. The material size distribution of corundum is given in Fig. 13.1.

10.1.2 Specimen Preparation

The objective of this method is mainly to improve the microtexture and macrotexture of ECC by mixing aggregates together. As the aggregates are mixed together with other material during the mixing process, it makes this process can be applied for both the "cast in situ" method and "prefabricating" method. Two aggregates are investigated: corundum and silica sand. ECC mixed with corundum is named ECC-Cor, while ECC mixed with silica sand is named ECC-SS. Both specimens are shown in Fig. 7.2.

ECC and ECC-Cor mixture was prepared in a planetary mixer. Cement, fly ash, and micro-silica sand are dried mixed first. Water was then added, followed by the addition of superplasticizer and mixed for 3–5 min until the mix achieved the required consistency. Lastly, the PVA fiber and corundum were added and mixed for another 3–5 min. The fresh mixture was cast into a prism (280 mm by 70 mm by 50 mm) and plate (500 mm by 300 mm by 10 mm) moulds catered for various types of tests. Specimens were removed from the mould after one day and were cured at normal laboratory air (25 ± 3 °C, 70%RH) for another 27 days before further testing. At the pre-determined age, either one day or three days, the top surface of plate specimens was treated with a steel brush to expose the corundum on the surface.

The second mixture, ECC-SS, is designed with multiple objectives. It offers good skid resistance with a brighter surface layer for lesser impact to UHI and better visualization. Thus, fine aggregate, silica sand, with high hardness (slightly below corundum) and bright color, is used. The ECC-SS mixture was prepared in a similar way to ECC-Cor.

The specimens were prepared for two different tests: outflow and water film thickness. The outflow test was done on the flat and grooved specimens. Both have a dimension of 500 × 300 × 10 mm. The flat specimens were reverse cast to ensure the surface is flat. The grooved specimens were also made with a reverse method. The grooves were successfully implemented by using a concrete cutter machine. The concrete cutter was equipped with a blade of 350 mm in diameter and 3.2 mm in thickness. The blade position is adjustable, so it can slide forwards and backwards, while the level can be fixed to make a groove with the specified depth. Four macro-textures were evaluated, and the corresponding groove dimensions are summarized in Table 37. Five minimum tests were conducted for each sample (see Fig. 10.1). As for the water film thickness test, flat and grooved specimens were also prepared but with different dimension. It has a dimension of 2 × 0.5 × 0.07 m in thickness.

(a) (b)

Fig. 10.1 Grooves preparation with **a** concrete cutter **b** silicone mould

Both specimens were cast with a reverse method. For the grooved ECC-Cor specimens, the grooves were made by using a rubber silicone mat as a negative mould (see Fig. 10.1).

10.1.3 Test Program

Outflow meter tests were conducted to evaluate the surface water drainage performance of the pavement specimens made of modified ECC through its surface and subsurface voids. It measures the duration of a known quantity of water taken to run through voids in the pavement texture and intended to provide a measure of the ability of the pavement to relieve pressure from the surface of the tires, thus indicating hydroplaning potential under wet conditions. A faster run time indicates a thinner film of water may exist between the tire (ASTM E2380M-15 2015). The duration measured with a digital timer with an accuracy of 0.00 s. Minimum five tests were conducted with a water temperature of 28 °C in an outdoor environment (air temperature of 30 °C). It is important to ensure the sample level is zero to avoid the effect of pavement geometry, the cross slope and the grades. The grade and the cross slope of the specimens were measured using a digital inclinometer (Fig. 10.2).

Water film thickness or water depth above the top of the specimen surface was also measured. Water depth is affected by many variables, including texture, pavement geometry, and environmental conditions [4]. Water depth has a strong correlation with hydroplaning and skid resistance. Therefore, it must be minimized (Fig. 10.3).

The equation of water depth [4] shown as:

$$\mathrm{WD} = 0.01485\left(\mathrm{TXD}0.11^{\mathrm{L_f}^{0.43}}\mathrm{I}0.59^{\mathrm{S_r}^{-0.42}}\right) - \mathrm{TXD} \tag{10.32}$$

where:

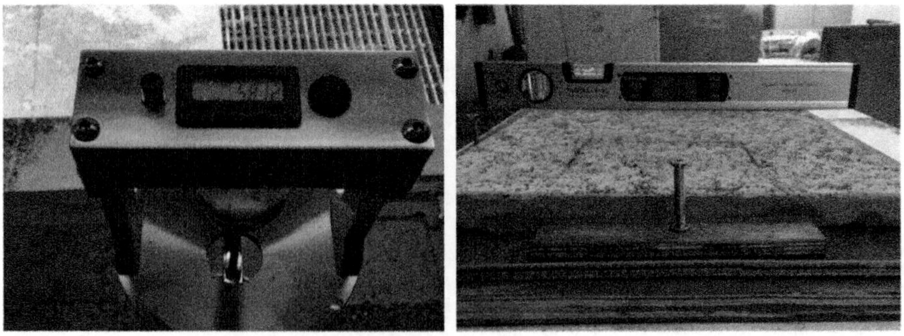

(a) (b)

Fig. 10.2 Water outflow measurement **a** with digital timer, **b** specimen level

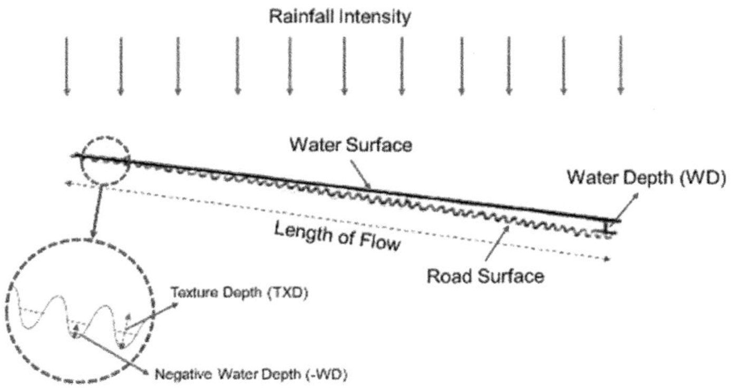

Fig. 10.3 Water depth (water film thickness) above the specimen surface [4]

WD = the water depth above the top of the surface asperities, mm
TF= length of flow path, m
TXD= texture depth, mm
I= average rainfall intensity (mm/hr)
S_r= slope of resultant flow path, m/m

In this study, the water depth is measured based on the difference of water volume between the input and the output within the system. The system is illustrated in the figure below. The water will be flowing from the inbox (input of the system) and then going on top of the specimen. Then the water will be caught again in the outbox. The water pump placed in the outbox will pump back the water again into the inbox via pipes. The flow rate of the water can be controlled with the valve as well as monitored with the flow meter (Fig. 10.4).

According to the theory mentioned above, a tailored apparatus was designed and made accordingly to do the investigation. The investigation on the water depth was done on a different flow rate of water and grade of the slab specimen. The flow

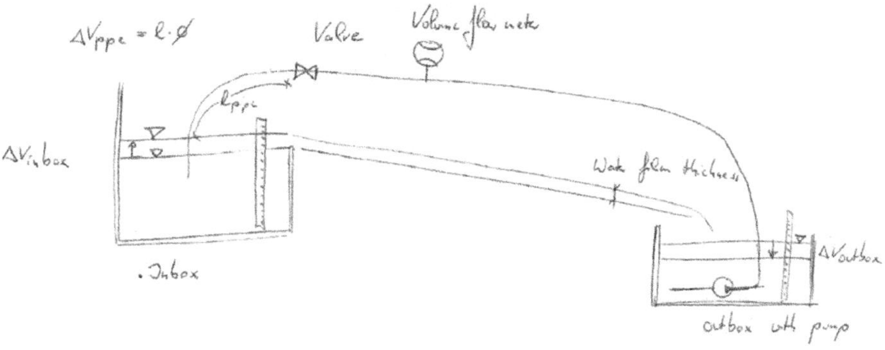

Fig. 10.4 The average water depth above the surface of specimen (Lechner B. & Bawono A. A)

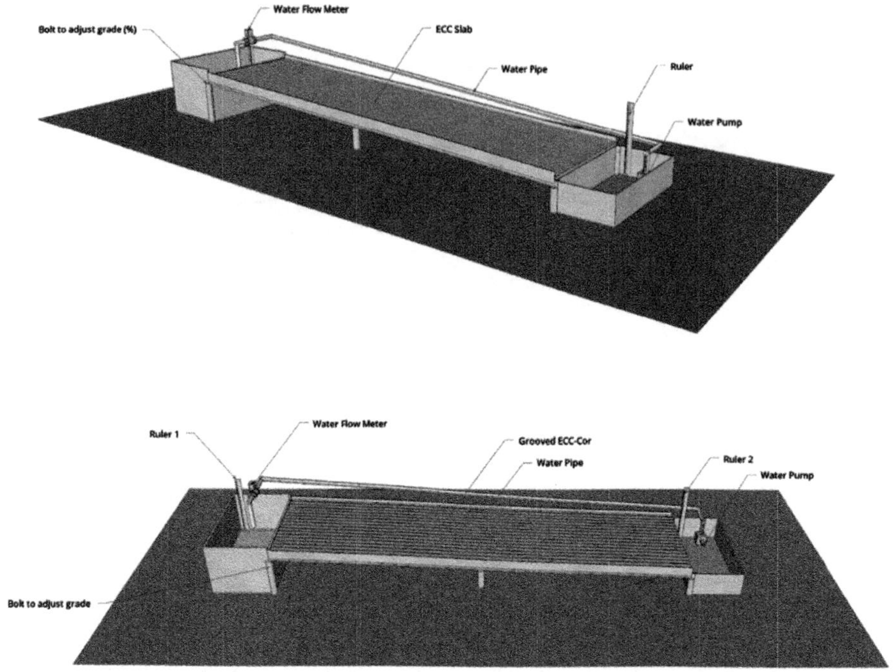

Fig. 10.5 The 3D illustration of the surface water drainage test setup **a** on ECC-Cor **b** on grooved ECC-Cor

rate of the water is varying: 1, 3, 5 and 9 L/minute. While two grades: 1 and 5%, are selected. The 1% grade is commonly used as the minimum grade in pavement geometry design. Meanwhile, a grade of 5% is considered as the maximum. Even though some geometry pavement guidelines state that grade more than 5% can be used in specific road segments (Fig. 10.5).

The equation of the average water depth can be seen below. The proposed experiment, as well as the equation, gives a different perspective in comparison to the method developed by Gallaway et al. The proposed equation gives the average water depth on the whole surface of the specimen, while the previous equation gives the water depth above the surface only at the edge of the specimen.

$$WD_{average} = \frac{\Delta V_{outbox} - \Delta V_{inbox} - \Delta V_{pipe}}{Area_{pavement}} \tag{10.33}$$

where:

$WD_{average}$=Water depth or water film thickness
V= Water Volume

Table 10.1 Laboratory test on the surface water drainage for surface treatments

No	Specimen	Water outflow	Water film thickness
1	ECC	Y	
2	Brushed ECC-Cor	Y	Y
3	Brushed ECC-SS	Y	
4	Grooved ECC-Cor	Y	Y

The intrinsic water permeability of a typical structural concrete is very low; it lies within a range of 5×10^{-19} to 2×10^{-17} m/s [11]. The water permeability of ECC is even lower compared to conventional concrete. Previous research was done on the permeability of ECC material by using two experimental setups: a falling head test which was used for specimens with low permeability, and a constant head test which was used for samples with higher permeability. Both tests done with the tensile specimen edges were sealed with epoxy to facilitate unidirectional flow. As results of the experimental tests, ECC M45 produced micro-cracks with crack widths ranging from 40–70 μm, and in the loaded state, the results are lower than the limit of 100 μm. In the unloaded scenario, ECC micro-cracks are observed to have crack widths from 5–10 μm. This led to water permeability of approximately 8.18×10^{-12} m/s and 7.74×10^{-10} m/s for uncracked ECC and ECC subjected to 3% strain, respectively. This means the water permeability remained very low for both conditions: cracked and uncracked. Water permeability does not change significantly when the number of cracks increases because it is maintained by the micro-cracks [6]. Therefore, water infiltration was ignored in this research and focused on the water surface runoff.

As can be seen from Table 10.1, various methods of surface treatments have been studied in this study. Each surface treatment is evaluated with an outflow test. Then, it is followed by a test on the water film thickness for specimens of ECC with corundum. As a reference, the outflow test was done on normal ECC samples.

10.2 Results

10.3 Water Outflow Performance of ECC, ECC-Cor, and ECC-SS

The outflow times on normal ECC cannot be measured as the surface is too smooth, that the rubber ring of the outflow meter device is stick to the surface tightly. This condition makes the water flowing out at a very slow rate. As shown in Fig. 10.6, the outflow times on ECC mixed with corundum are improved as the corundum provides a microtexture, allowing the water to find a gap between the aggregates. The average outflow time result of ECC-Cor is 62.11 ± 4.99 s. A similar result is observed in ECC-SS 01 with 65.06 ± 3.51 s.

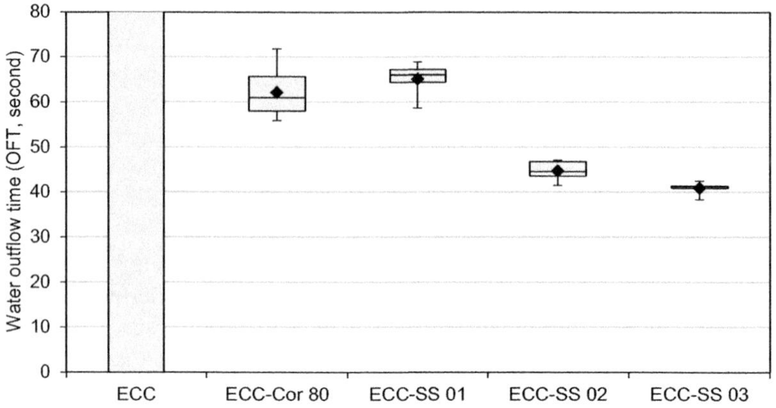

Fig. 10.6 OFT of ECC, ECC-SS #, and ECC-Cor 80

10.3.1 Water Outflow Performance on Grooved ECC-Cor

As expected, the outflow times for the grooved ECC-Cor are even shorter. The combination of the microtexture provided by the corundum and the macrotexture provided by the grooves makes the space between the rubber ring of the Outflow Meter device and the surface pavement specimens even bigger, which results in a lower of outflow times. The outflow times of grooved ECC-Cor with a land area of 25.0 ± 0.5, 20.0 ± 0.5, 15.0 ± 0.5, and 7.5 ± 0.5 mm is 7.73, 7.30, 5.8 and 3.38 s, respectively. Detail results are shown in Fig. 10.7.

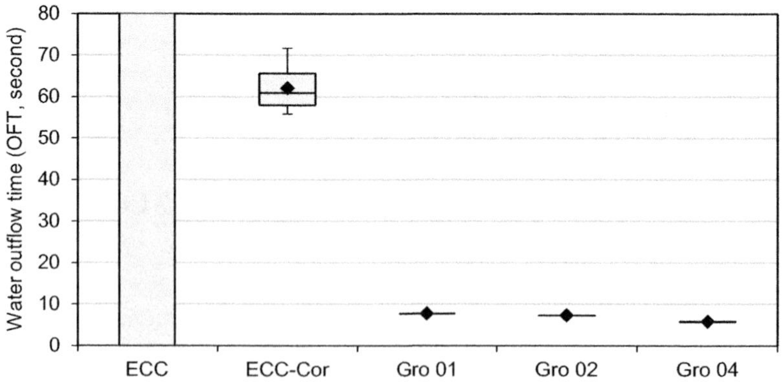

Fig. 10.7 OFT of ECC, ECC-Cor 80, ECC-Cor Gro #

10.3.2 Water Film Thickness on ECC-Cor

It can be seen in Fig. 10.8, the average WFT on ECC-Cor for water drained above the ECC-Cor surface pavement (grade 1%) with a flow rate of 1.0, 3.1, 5.1, and 8.9 L/minute, are 1.33, 1.91, 2.66, and 4.10 mm, respectively. It can be seen in Fig. 10.9, the average WFT on ECC-Cor for water drained above the ECC-Cor surface pavement (grade 5%) with a flow rate of 1.0, 3.1, 5.1, and 8.9 L/minute, are 1.03, 1.18, 1.93, and 2.78 mm, respectively. It is not much difference between the average WFT results derived from the equation and direct measurement (<10%).

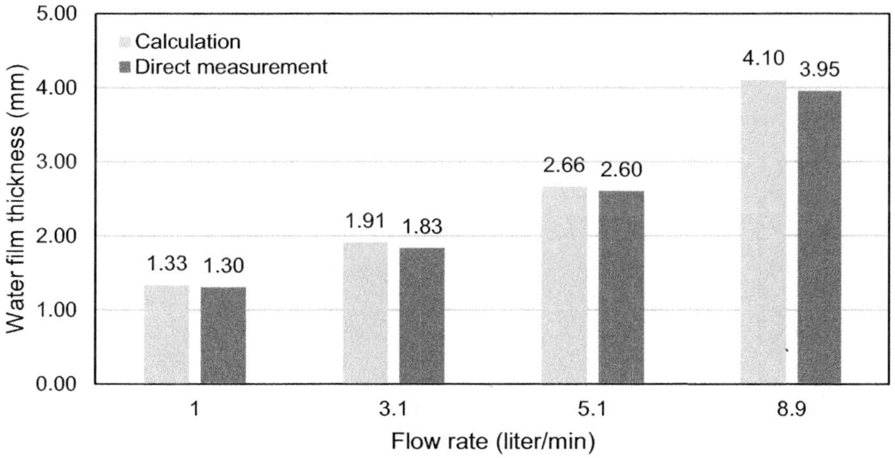

Fig. 10.8 Average WFT on flat ECC surface (grade 1%)

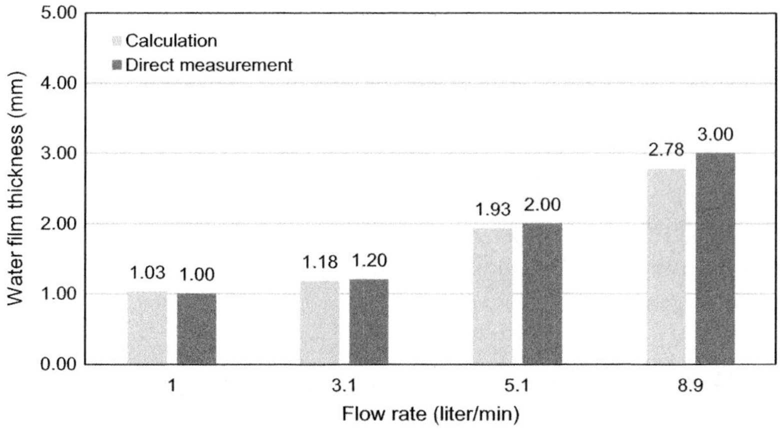

Fig. 10.9 Average WFT on flat ECC surface (grade 5%)

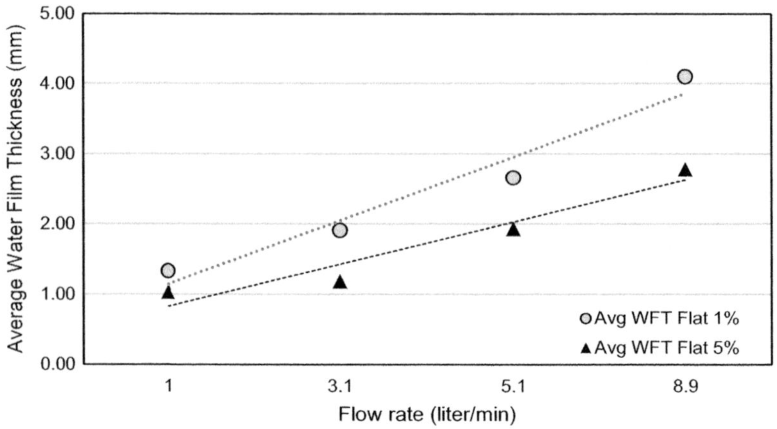

Fig. 10.10 Average WFT on flat ECC surface (grade 1% vs. 5%)

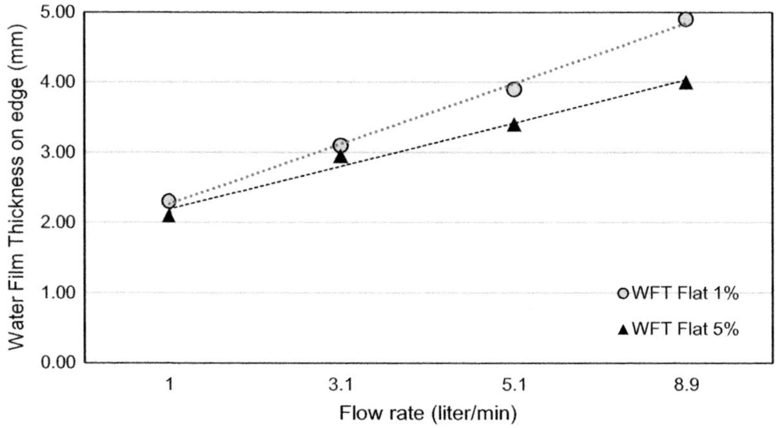

Fig. 10.11 Average WFT on edge for flat ECC surface (grade 1% vs. 5%)

The difference between the average WFT of ECC-Cor with a grade of 1% and 5% are shown in Fig. 13.2. Meanwhile, the WFT on the edge of the pavement can be seen in Fig. 10.11. As depicted from these figures, when the grade of the slab increased, the average WFT will be reduced (Fig. 10.10).

10.3.3 Water Film Thickness on Grooved ECC-Cor

It can be seen in Fig. 10.12, the average WFT on grooved ECC-Cor for water drained above the ECC-Cor surface pavement (grade 1%) with a flow rate of 1.0, 3.1, 5.1,

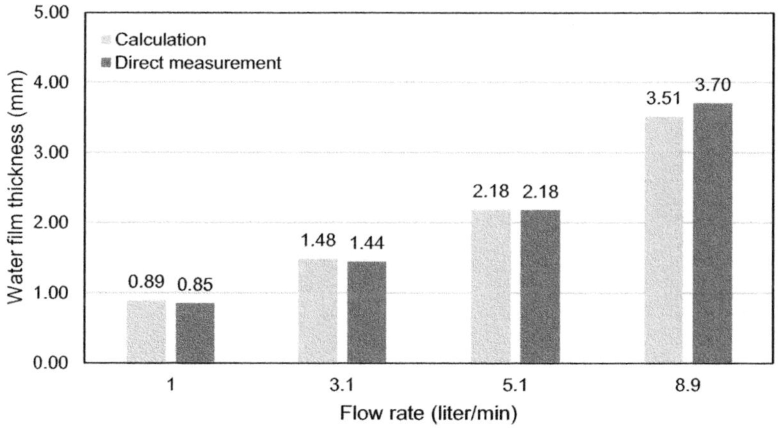

Fig. 10.12 Average WFT on grooved ECC surface (grade 1%)

and 8.9 L/minute, are 0.09, 0.15, 0.22, and 0.35 mm, respectively. It can be seen in Fig. 10.13, the average WFT on grooved ECC-Cor for water drained above the ECC-Cor surface pavement (grade 5%) with a flow rate of 1.0, 3.1, 5.1, and 8.9 L/minute, are 1.03, 1.18, 1.93, and 2.78 mm, respectively. It is not much difference between the average WFT results derived from the equation and direct measurement (<10%).

The difference between the average WFT of ECC-Cor with a grade of 1 and 5% are shown in Fig. 10.14. The WFT on the edge of the pavement can be seen in Fig. 10.15. As depicted from these figures, when the grade of the slab increased, the average WFT will be reduced.

10.3.4 Water Film Thickness Comparison Between Flat and Grooved ECC-Cor

The WFT will be significantly decreased when such a surface pavement equipped with grooves. The comparison results between the flat and grooved ECC-Cor specimens can be seen in Figs. 10.16 and 10.17 for average WFT and WFT on edge, respectively.

In summary, conclusions can be drawn, including:

- The WFT will be significantly decreased when such a surface pavement equipped with grooves, as the water will fill the grooves immediately when drained above the surface. Once the grooves are filled with water, the remaining water will be drained along the pavement's surface.
- The grooves can reduce the WFT up to 50% of its original thickness, depends on the dimension of the grooves, as well as the grade of the pavement.

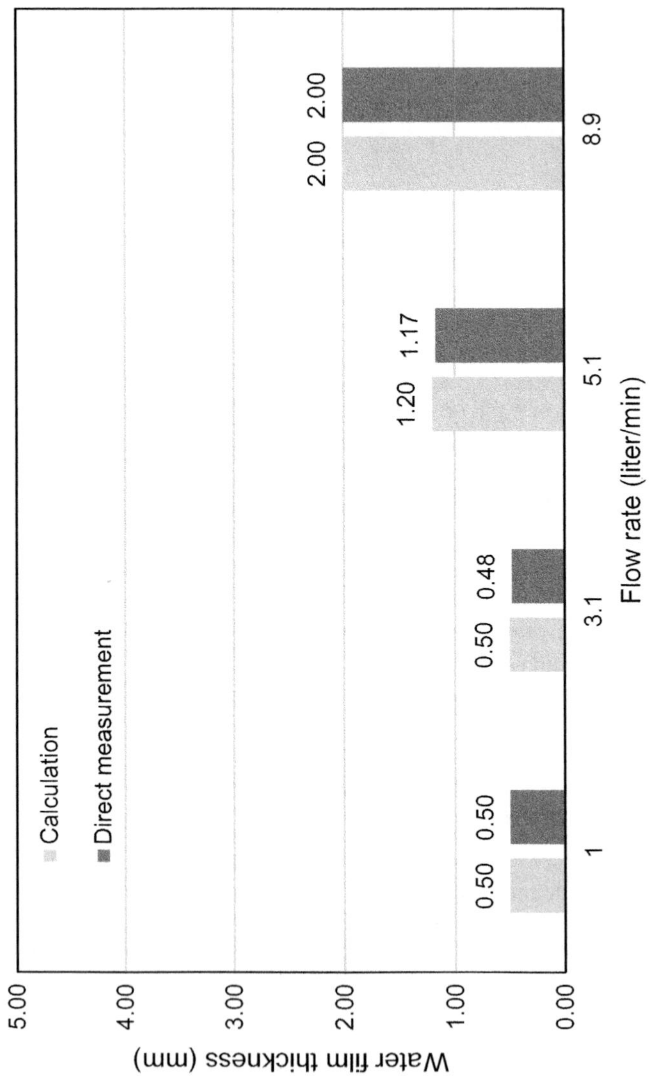

Fig. 10.13 Average WFT on grooved ECC surface (grade 5%)

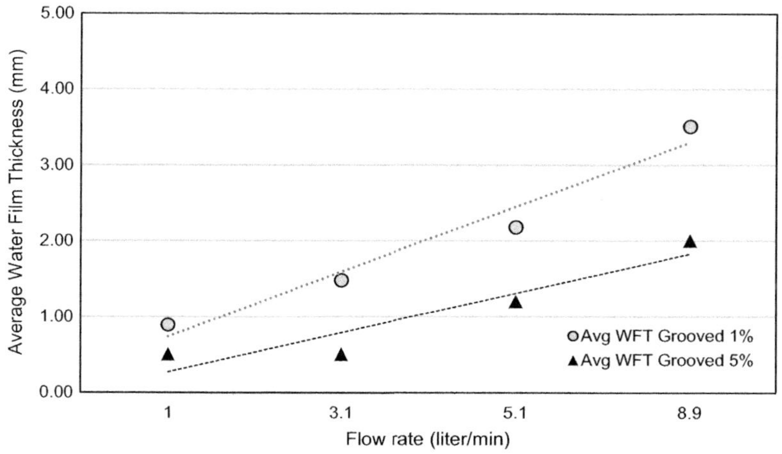

Fig. 10.14 Average WFT for grooved ECC surface (grade 1% vs. 5%)

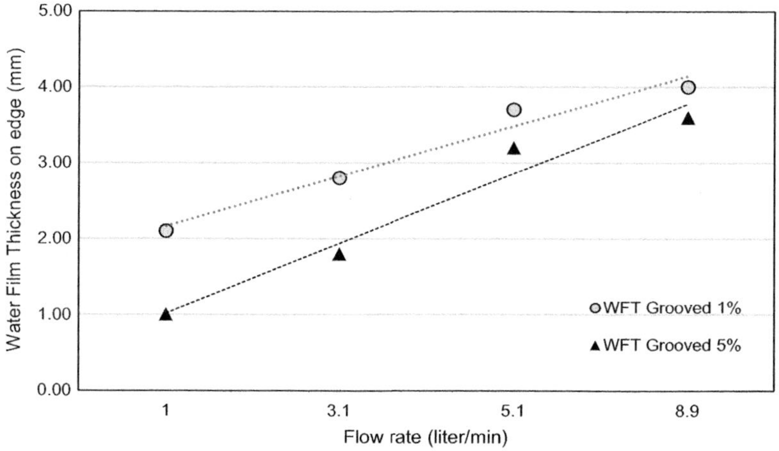

Fig. 10.15 WFT on edge for grooved ECC (grade 1% vs. 5%)

- The higher the grade, will result in the reduction of WFT. With a similar dimension of grooves, the WFT is reduced up to 30% between the flat and grooved ECC-Cor with a 1% grade. In comparison, the WFT can be reduced up to 50% between the flat and grooved ECC-Cor with a 5% grade.

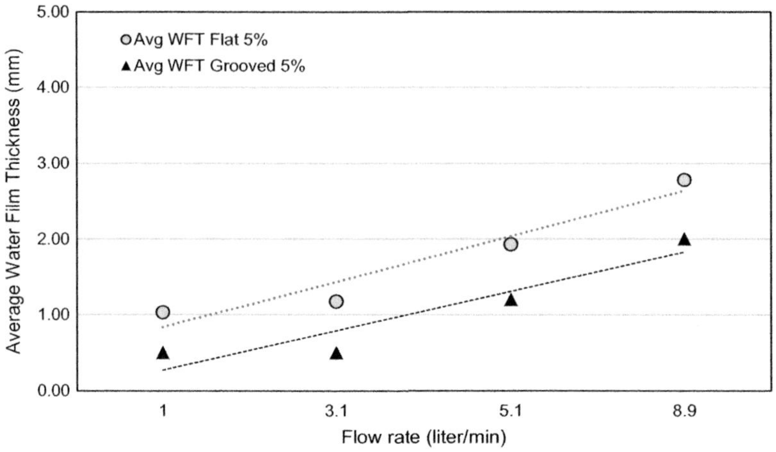

Fig. 10.16 Average on WFT on flat vs. grooved ECC-Cor (grade 5%)

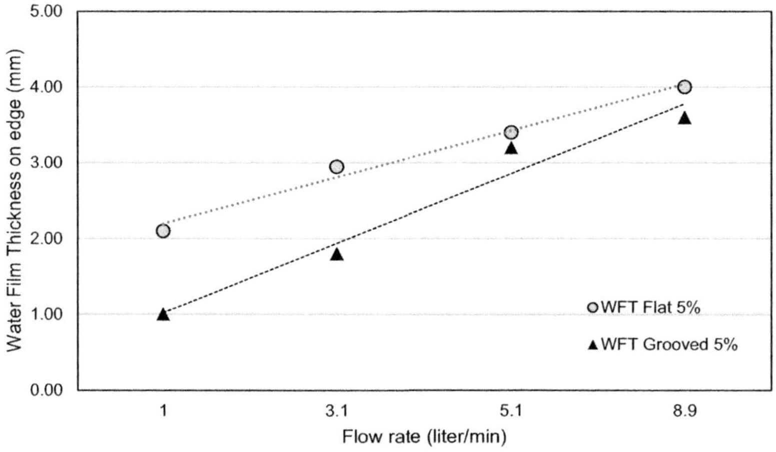

Fig. 10.17 WFT on edge for flat vs. grooved ECC-Cor (grade 5%)

References

1. ASTM E2380M-15 (2015) Test method for measuring pavement texture drainage using an outflow meter. ASTM International, West Conshohocken, PA
2. Browne A (1971). Dynamic hydroplaning of pneumatic tires. Northwestern University
3. Current Results Publishing Ltd. (2016) Average annual precipitation by state. https://www.currentresults.com/Weather/US/average-annual-state-precipitation.php. Accessed Oct 2016
4. Gallaway B, Ive D, Hayes G, Ledbetter W, Olson R, Woods D, Schiller Jr R (1979) Pavement and geometric design criteria for minimizing hydroplaning. FHWA-RD-79–31
5. Gallawa BM, Schiller RE, Rose JG (1971) Effects of rainfall intensity, pavement cross slope, surface texture, and drainage length on pavement water depths. Research report 138-5, Texas

6. Lepech MD, Li VC (2009) Water permeability of engineered cementitious composites. Cement and concrete composites, 31(1 ASTM E2380M-15 (2015). Test method for measuring pavement texture drainage using an outflow meter. ASTM International, West Conshohocken, PA
7. Moore DF (1967) A theory of viscous hydroplaning. Int J Mech Sci 9:797–810
8. PUB Singapore (2013) Managing urban runoff (Drainage handbook). PUB and IES, Singapore
9. Road Design Drainage Technical Sub-committee (2013) Guide to road design part 5a: drainage—road surface, networks, basins, and subsurfaces. Austroads, Sydney
10. Tubey L, Hosking J (1972) Synthetic aggregates of high resistance to polishing—corundum rich aggregates. TRRL LR467
11. Villar MV, Martin PL, Romero FJ, Gutierrez-Rodrigo V, Barcala JM (2015) Gas and water permeability of concrete. Geological Society, London, Special Publications 415(1):59–73. https://doi.org/10.1144/SP415.6

Chapter 11
Study of Surface Water Drainage: Field Investigation

11.1 Introduction

This chapter presents the investigation of surface water drainage on the full-scale application of modified ECC for pavement application. Data from various texturing methods applied in the field have been investigated and then analyzed. The test includes water drainage performance with an outflow meter. Different texturing methods applied to find the optimum solution to produce good surface properties for pavement application.

In comparison to a laboratory scale, more variables need to be considered for a full-scale application of modified ECC, i.e., temperatures, mixing method, and finishing work. Test results show that the surface water drainage of the surface pavement can be improved by adopting hard fine aggregates into the ECC mixture, as it provides both the microtexture and macrotexture. It is also shown that texturing methods which was studied on laboratory scale tests can be further applied to full-scale application with consistent results.

11.2 Full-Scale Field Investigation

11.2.1 Test Program

Water outflow tests have been performed on the field to investigate surface water drainage performance on full-scale application [2, 3]. Many ideal conditions which can be found in the laboratory is simply not existing in the field. The type of mixer, duration of mixing, temperature, casting process can be the factors that simply affect the results compare to results acquired in a laboratory test scale. Various methods of surface treatments investigation in full-scale are summarized in Table 11.1.

The first method, brushed ECC mixed with aggregates, was done using the same method mentioned in 8.2.1. Meanwhile, the second, third, and fourth methods are

A. A. Bawono, *Engineered Cementitious Composites for Electrified Roadway in Megacities*, Springer Theses, https://doi.org/10.1007/978-3-030-88542-7_11

Table 11.1 Site Investigation on surface water drainage based on surface treatments

No	Surface treatment method	Precast	Cast in place	Reference
1	Brushed ECC mixed with aggregates	–	Y	Sect. 8.2.1
2	Exposed aggregates into ECC surface	Y	Y	Sect. 8.2.2
3	Prefabricating modified ECC slabs with steel formwork	Y	–	Sect. 8.2.4
4	Reverse prefabricating slabs of ECC mixed with aggregates	Y	–	Sect. 8.2.5

done according to the same method mentioned in Sect. 8.2.2, 8.2.4, and 8.2.5, respectively.

Water outflow tests were conducted to evaluate the drainage performance of the pavement through its surface and subsurface voids. It measures the duration of a known quantity of water taken to run through voids in the pavement texture and intended to provide a measure of the ability of the pavement to relieve pressure from the surface of the tires, thus indicating hydroplaning potential under wet conditions. A faster run time indicates a thinner film of water may exist between the tire [1]. The duration measured with a digital timer with an accuracy of 0.00 s. Minimum five tests were conducted with water with a temperature of 28 °C in an outdoor environment (air temperature of 30 °C). The grade and the cross slope of the specimens were measured using a digital inclinometer.

11.3 Results

11.3.1 Water Outflow of Brushed ECC-Cor

As can be seen from Fig. 11.1, the water outflow test results of brushed ECC mixed with corundum improved when there is more corundum on the surface as they have been exposed (see after 6 and 21 h). When the matrix is not fully hardened, more corundum aggregated have been removed away in the process. This phenomenon can be seen in the "after four hours exposed" part.

The difference due to the processing time does not affect water outflow results, as all of them are below 10 s. It can be concluded, based on microtexture and macrotexture, good results are coming from brushing time in the range between 6–12 h. However, it should be mentioned that the results from the field investigation are faster than results acquired in laboratory tests, where outflow times are approximately more than 60 s and less than 10 s for laboratory tests and field investigation results, respectively. This occurs because the levelling process of ECC in a full-scale application is considerably difficult, resulting in a rougher surface.

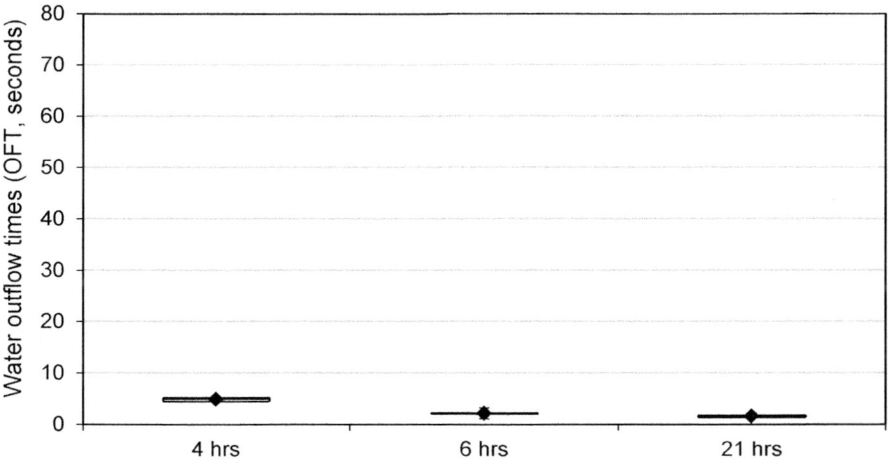

Fig. 11.1 OFT of brushed ECC-Cor with cast in situ method

11.3.2 Water Outflow Performance of Exposed ECC-Cor

OFT results for the exposed aggregates on the ECC method, which is applied for slab and footpath application, are shown in Fig. 11.2. Additional results for concrete M45, which was applied on a footpath, is also added for comparison. In comparison to the concrete surface with tining surfacing method, the modified ECC with exposed aggregates (both for footpath and slab) results in faster water outflow (<10 s). Thus, exposed aggregates on the ECC surface shown as a good potential method to improve the surface water drainage.

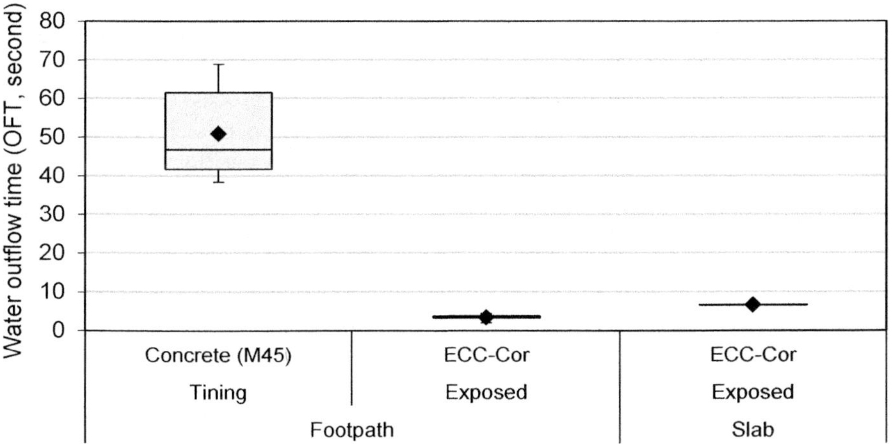

Fig. 11.2 OFT of exposed aggregates on ECC

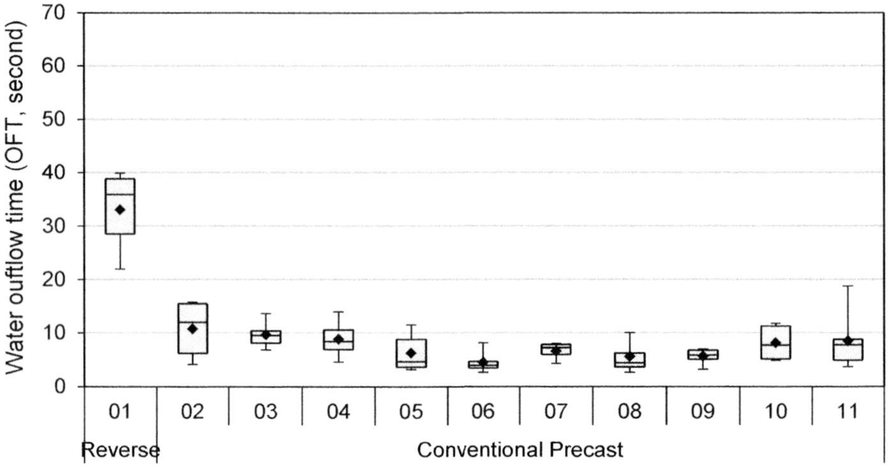

Fig. 11.3 OFT of precast and reverse precast ECC-Cor (busbay application)

11.3.3 Water Outflow Performance of Pavement Made of ECC-Cor with Precast Method

OFT tests were taken on-site at the bus stop for slabs made with the precast method, both with reverse and conventional precast. According to the results shown in Fig. 11.3, ECC-Cor with reverse precast method (method 4th), which is slab 01, results in a slower rate of outflow times compared to the conventional precast method (slab 02–11). This occurred because the ECC is made with a reverse precast method that produces a flat surface compared to the conventional precast method. As has been mentioned, the flat state of the surface leads to a lesser gap between the surface of ECC and the rubber of the outflow meter.

Various results can be seen from ECC slabs made with the conventional precast method (slab no 02–11). The results range between 3–19 s, with an average of 7 s, which again may happen due to an unstandardized casting process. Less variety can be expected when slabs can be manufactured with more proper machines and human resources. The number of the slab is in accordance with the plan illustrated in Fig. 8.9.

In summary, the brushed method can be used as a surface treatment for pavement made of ECC if the surface can be produced flatly. It is important to do the brushing for not more than 24 h to avoid the cementitious material is becoming completely hardened. Thus, it sticks to the surface while the surface of the pavement is expected to have some texture.

From the same results, another conclusion can be drawn that the reverse precast method produces a much flatter surface compared to the conventional precast method. However, the brushing method produces a slower outflow rate compared to the exposing method.

References

1. ASTM E2380M-15 (2015). Test method for measuring pavement texture drainage using an outflow meter. ASTM International, West Conshohocken, PA
2. Browne A (1971) Dynamic hydroplaning of pneumatic tires. Northwestern University
3. Gallaway B, Ivey D, Hayes G, Ledbette W, Olso R, Woods D, Schiller Jr R (1979) Pavement and geometric design criteria for minimizing hydroplaning. FHWA-RD-79–31

Chapter 12
Study of Surface Water Drainage: Numerical 3D CFX

This chapter describes the numerical study employed by using ANSYS Fluent (Fluid Dynamics Analysis) developed in this thesis. The numerical study consists of the development of the 3D model for the simulation of surface water drainage. The 3D model includes geometry development, mesh design, material properties, conditions, model validation and analysis. The objective of the model is to simulate how the surface texture and geometry of pavement made of ECC will affect water surface runoff.

It has been highlighted that doing an analysis using a numerical model on water surface runoff will be helpful to study the rainfall loading on the highway in Singapore. Various textures, cross falls, and slopes that affect the surface drainage performance of the pavement can be studied by using fluid dynamics. However, until now, there is no numerical model on water surface runoff in Singapore condition where rainfall loading is considerably high. Such a study will be quite useful to determine the surface properties of pavement design for SRT.

Computational Fluid Dynamics (CFD) is a software-based tool used for simulating behavior such as fluid flow, heat transfer, and other related physical processes. It works by solving the equations of fluid flow over a region with a designed boundary. The equation is in conjunction with the Navier–Stokes. There are methods that can be used to solve the CFD model. However, the most common one is the finite volume technique or also known as the Finite Volume Model (FVM). In this FVM, the region of interest will be divided into small sub-regions, called control volumes. The equations are discretized and solved iteratively for each control volume. As results, an approximation of the value of each variable at specific points throughout the domain can be obtained [1].

ANSYS CFX provides comprehensive modelling capabilities for CFD analysis with a wide range of incompressible, compressible, laminar and turbulent fluid flow problem solutions. Steady-state or transient analyses can be performed. In ANSYS CFX, a broad range of mathematical models for transport phenomena (like water transfer) is combined with the ability to model complex geometries.

A. A. Bawono, *Engineered Cementitious Composites for Electrified Roadway in Megacities*, Springer Theses, https://doi.org/10.1007/978-3-030-88542-7_12

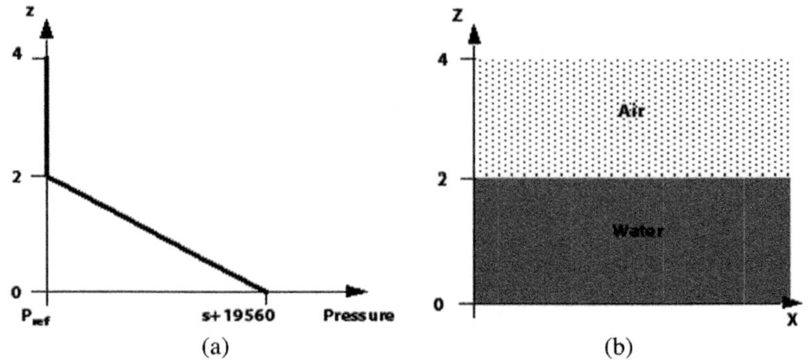

Fig. 12.1 Liquid phase of fluid dynamic model [1]

Simulation of water transport can be done using the Volume of Fluid (VOF) model. The VOF model can model two or more immiscible fluids by solving a single set of momentum equations and tracking the volume fraction of each of the fluids throughout the domain. ANSYS CFX can model the effects of open channel flow (for example, rivers, dams, and surface-piercing structures in an unbounded stream) using the VOF formulation and the open channel boundary condition. These flows involve the existence of a free surface between the flowing fluid and fluid above it (generally the atmosphere). In such cases, wave propagation and free surface behavior become important. Flow is generally governed by the forces of gravity and inertia. This feature is mostly applicable to marine applications and the analysis of flows through drainage systems.

Free surface flow refers to a multiphase flow situation where a distinct interface separates the phases. Examples of free surface flows include open channel flow, flow around ship hulls, tank filling problems, Pelton turbines, and many others. Free surface flow with the homogeneous model should be used when possible.

Free surface flow model criteria:

- For the Buoyancy Reference Density, choose the density of the light fluid (air). This simplifies the definition of pressure, and hence the specification of pressure initial and boundary conditions. In the hydrostatic limit:
- At pressure outlets, the elevation of the water must usually be known as part of the problem definition. The mechanism used to enforce this elevation is to set a pressure profile consistent with the known elevation. In some circumstances, the elevation is not known in advance. In this case, it is best to specify a pressure profile consistent with an estimate for the outlet elevation. This assumption should not affect the flow other than possibly inducing a small kink in the flow just upstream of the outlet condition.
- The initial conditions for the pressure field and volume fraction must be consistent (that is, the pressure field is hydrostatic in the heavy phase and uniform in the light phase). This is easily done using CEL step functions. For example, the functions (Fig. 12.1).

Table 12.1 Material properties for CFX model on surface water drainage

Material	Water	ECC
Elastic modulus (MPa)	N.A	22.000
Poisson ration	N.A	0.15
Density (kg/m^3)	997	2100

12.1 Material Properties

Material properties are determined in the early stage of ANSYS development. Materials used for the CFX model includes ECC for the sample tested, water and air. Those are the inputs in the Engineering Data module in ANSYS software. ECC has an Elastic Modulus of 18 to 34 GPa with Poisson's ratio of a range from 0.15 to 0.20 [4, 6]. In this 3D FEM model, the value of 22 GPa and 0.15 were adopted for elastic modulus and Poisson's ratio, respectively. The densities of ECC samples were calculated: 2040 and 2190 kg/m^3 for the ECC mixed corundum with Pour and Precast Method. Therefore, for this model, an average value of 2100 kg/m^3 has been used for the density of the specimen. This is complying with typical Portland cement mortar, which normally has a density of around 2000–2200 kg/m^3. However, in this model analysis, the density might not affect much for the results. Normal material properties for water and air have been used (Table 12.1).

12.2 Geometry Model

A 3D model of the surface water drainage test on the ECC surface was created with the DesignModeler tool in ANSYS. As can be seen in Fig. 12.2, the dimension of the ECC specimen is set accordingly to the actual measurement. The model consists of two parts: (1) the main body with a dimension of 2 m by 0.5 m by 0.05 m that is made of ECC material, which is where the water and air can flow on top of the surface. The angle of the body can be adjusted according to needs. In this model, it was designed for an angle of 1° and 5°. As shown in Fig. 12.2b, the main body angle is adjusted to an angle of 1°, (2) the second body is attached to the main body, as it was designed as the output of the water from the main body. The second body is fixed (its angle cannot be adjusted).

12.3 Mesh Design

The main objective of this 3D CFX model is to simulate the surface water drainage performance on the ECC sample. The meshing was created with the help of the Sweep Method and Face Sizing tools to create a rectangular shape with a maximum element

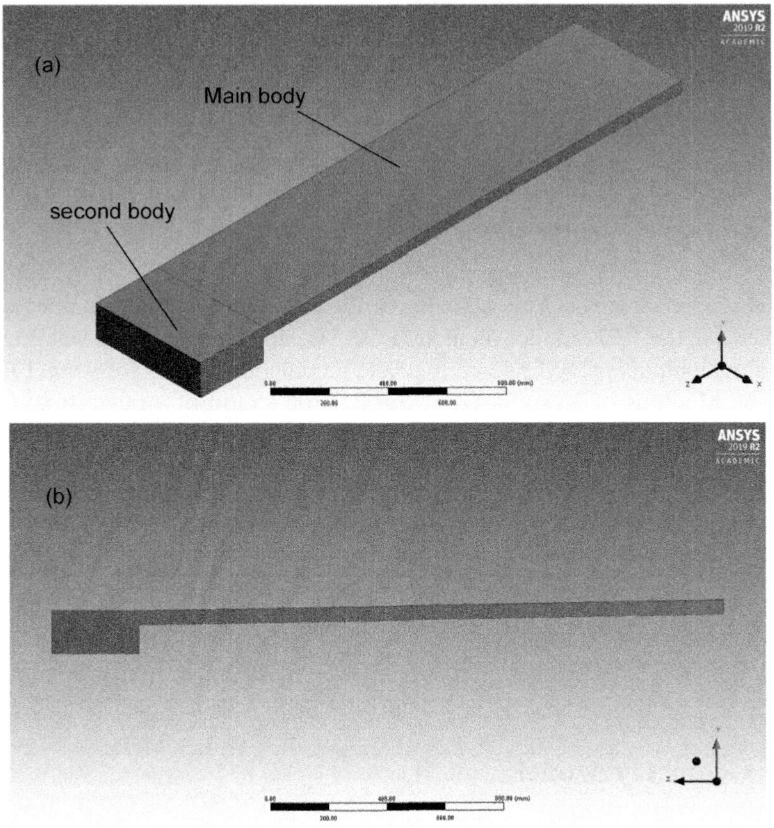

Fig. 12.2 3D CFX model on surface water drainage, **a** bird view, **b** side view

size of 0.5 mm. The meshing comprised 1,589,367 nodes and 375,000 elements (Fig. 12.3).

The quality of the mesh has been checked in the ANSYS program. ANSYS recommend keeping minimum orthogonal quality above 0.1. Another parameter is the aspect ratio, which can be calculated by the length divided by the height. This model has an aspect ratio of 5.99 as average. Even though the Element Quality and Aspect Ratio show the rather low-quality meshing, it is still acceptable as these values are generated from the average of the whole system. More importantly, it is the meshing where the water is flowing where the meshing was designed denser. The results for meshing are summarized in Table 12.2.

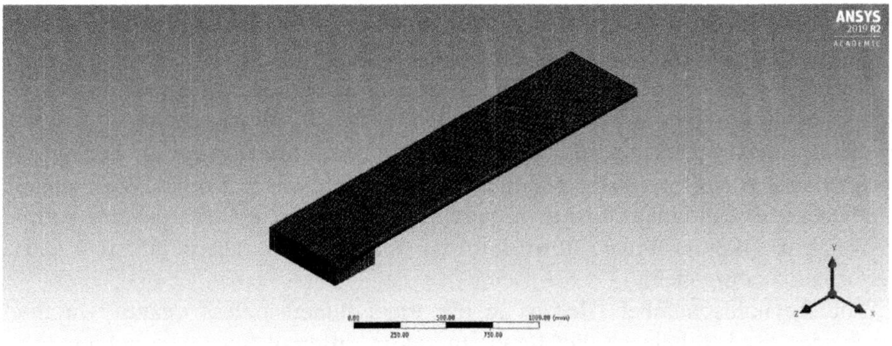

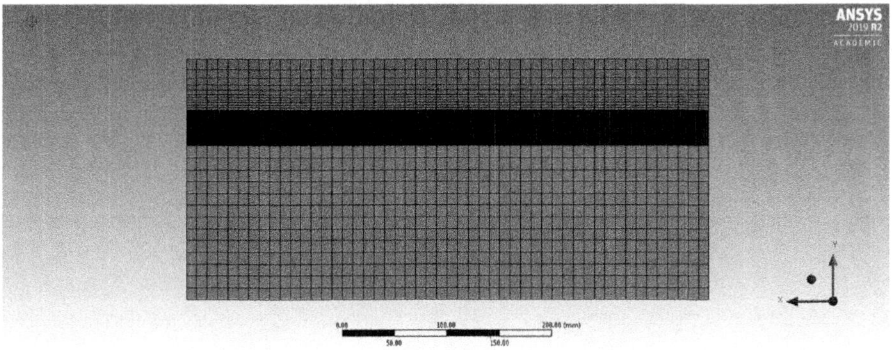

Fig. 12.3 Meshing design on 3D CFX model **a** from general view, **b** with denser mesh at "inlet"

Table 12.2 Quality of meshing for 3D FEM model of surface water drainage test

Criteria	Value	Standard criteria
Element quality	0.38 ± 0.21	0–1 0 is worst, 1 is best
Orthogonal quality	0.79 ± 0.17	0–1 1 indicates a perfect cube or square; while 0 indicates element has a zero or negative volume
Aspect ratio	5.99 ± 3.3	The best possible quadrilateral aspect ratio is 1

12.4 Surface Water Drainage 3D CFX Model

The 3D CFX surface water drainage model is designed based on the steady-state model. A flow that is not a function of time is called steady flow. Steady-state flow refers to the condition where the fluid properties at a point in the system do not change over time. Time-dependent flow is known as unsteady (also called transient). Whether a particular flow is steady or unsteady, it can depend on the chosen frame of reference. For instance, laminar flow over a sphere is steady in the frame of reference that is stationary with respect to the sphere. In a frame of reference that is stationary with respect to a background flow, the flow is unsteady.

Turbulence is flow characterized by recirculation, eddies, and apparent randomness. Flow in which turbulence is not exhibited is called laminar. It should be noted, however, that the presence of eddies or recirculation alone does not necessarily indicate turbulent flow—these phenomena may be present in laminar flow as well. Mathematically, turbulent flow is often represented via a Reynolds decomposition, in which the flow is broken down into the sum of an average component and a perturbation component.

The Reynolds number (Re) is an important dimensionless quantity in fluid mechanics used to help predict flow patterns in different fluid flow situations. It has wide applications, ranging from liquid flow in a pipe to the passage of air over an aircraft wing. With respect to laminar and turbulent flow regimes: laminar flow occurs at low Reynolds numbers, where viscous forces are dominant, and is characterized by smooth, constant fluid motion. While turbulent flow occurs at high Reynolds numbers and is dominated by inertial forces, which tend to produce chaotic eddies, vortices and other flow instabilities. The Reynolds number is defined as:

$$R_e = \frac{\rho u L}{\mu} = \frac{u L}{v} \tag{12.1}$$

where

- ρ density of the fluid (SI units: kg/m^3).
- u velocity of the fluid with respect to the object (m/s).
- L characteristic linear dimension (m).
- μ dynamic viscosity of the fluid (Pa s or N s/m^2 or kg/(m s)).
- ν kinematic viscosity of the fluid (m^2/s).

There are at least four boundaries that need to be set up in this model, namely as (1) inlet, (2) outlet, (3) free, and (4) ECC. Boundary needs to be set up as it functioned as artificial walls to prevent the water (fluid) is flowing from the system to outside. The inlet was set up as the "inlet", which produces a low intensity of turbulence. The volume fraction of water is 1, while air is 0. The outlet was set up as the "outlet" with Average Over Whole Outlet option to get the distributed output. The free-body was set up as the "opening", which produces a low intensity of turbulence. The volume fraction of air is 1, while water is 0. And then ECC was set up as the "wall" (Fig. 12.4).

12.5 Model Validation

The 3D CFX surface water drainage model has been calibrated and validated, including (1) roughness properties, (2) the calibration of correlation between WFT measurement in a laboratory test with WFT results generated from the CFX 3D model.

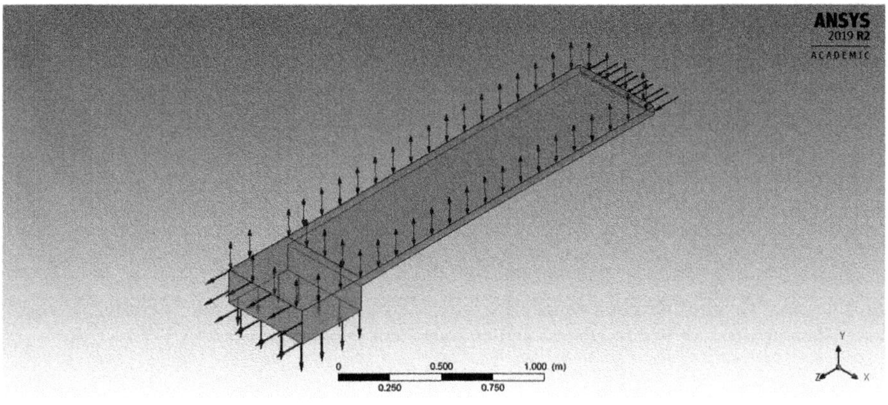

Fig. 12.4 Boundary set up in CFX model system

The ECC surface property in the 3D model has been calibrated accordingly as a rough wall with sand grain roughness set up based on the macrotexture property of ECC, which is 0.36 mm. This means the study considered the effect of surface roughness on the flow of the water. The roughness of the surface, which is set as a wall in the model, will add an additional term to the equation that reduces the mean velocity of the water along the wall [3, 5]. In this CFX model, the conversion from a real roughness height (K) of ECC material can be safely assumed to an equivalent roughness height (K_S), which then derived as ($K = K_S$) [2]. As results, the ECC wall is physically displaced in an upwards direction by $K_S/2$ (ANSYS) (Figs. 12.5 and 12.6).

Two full-scale of ECC specimens were built with a dimension of 2 m by 0.5 m by 0.1 m to validate the model. One ECC specimen is with a smooth surface, and the other one is equipped with grooves. Then the water flowing above the ECC surface with a steady flow rate. The angle of the ECC specimen is adjusted to a certain angle (1%). Then the water film thickness above the ECC specimen is measured with a minimum of nine measurements (three measurements on 1/3 length of the specimen,

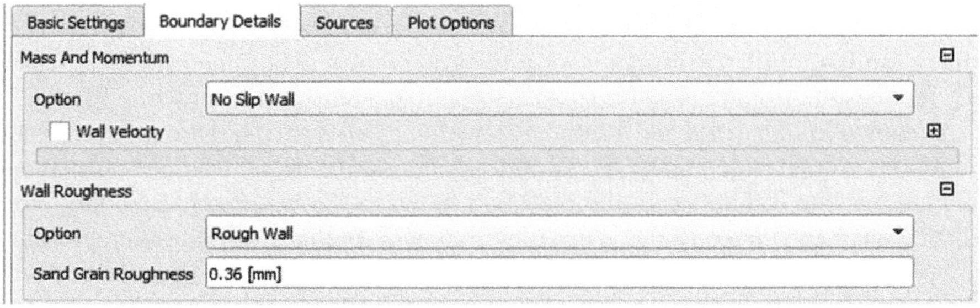

Fig. 12.5 Wall roughness in CFX model

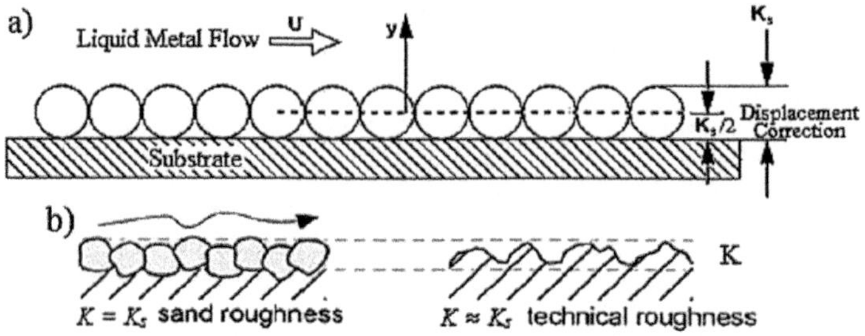

Fig. 12.6 Equivalent sand grain roughness **a** monolayer of spheres as rough layer in the CFX model, **b** comparison of real roughness and its equivalent sand roughness (ANSYS)

three measurements on ½ length of the specimen, and the rest three measurements on 2/3 length of the specimen). The average value is then taken and then used to validate the CFX model.

12.6 Results

The results are derived after the solution are converged through the simulation. Convergence control was set up with 2500 maximum iterations with residual target 0.0001, meaning the solution will stop and derived when the residual RMS reach this value. This value is taken as it is considered as a good result with a very minimum standard deviation (ANSYS).

Two methods are used to show the results of the 3D CFX model. Firstly, the results are shown with the isosurface method, where the surface of the water can be generated automatically. The input used for the isosurface is the water volume fraction with water velocity as a variable. Secondly, the results are generated by using the streamline or isoclip method, where several lines are represents the water's surface. The isosurface meant more for visualization, as it produces a better image and or video for the animation of the water above the ECC specimen itself. This method can be used for quick analysis, whether the model and results generated are properly working. Once the model is working, more detailed results can be generated with streamline. With streamline/isoclip, the water flow will be generated in the shape of lines. Results from the isosurface and streamline analysis are shown in Fig. 12.7.

As shown in Figs. 12.8 and 12.9, the elevation of the water and the concrete lines are plotted in the same chart. The elevation is a relative value. The only important parameter is the difference in the elevation between the water and concrete lines, which is then calculated as the water film thickness. In this study, the average value of the water (surface) lines is taken and considered as the main value, which will later be used as a comparison to the laboratory test results. Both results from 3D CFX model and laboratory measurement (including results from the proposed equation) are summarized in Table 12.3.

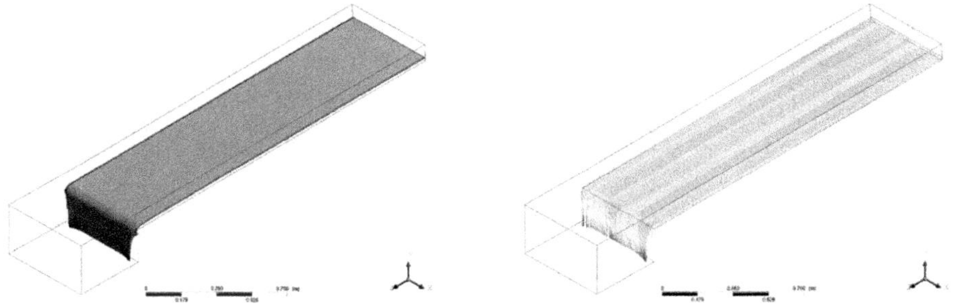

Fig. 12.7 CFX results of WFT on ECC-Cor with **a** isosurface, **b** isoclip

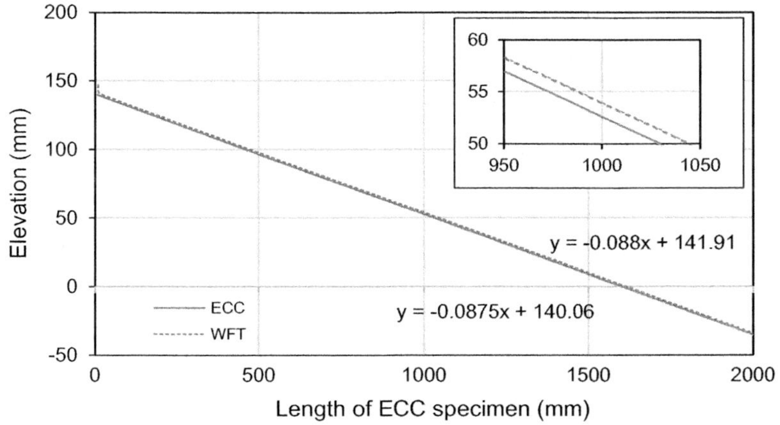

Fig. 12.8 Elevation of water and ECC-Cor based on isoclip method with CFX

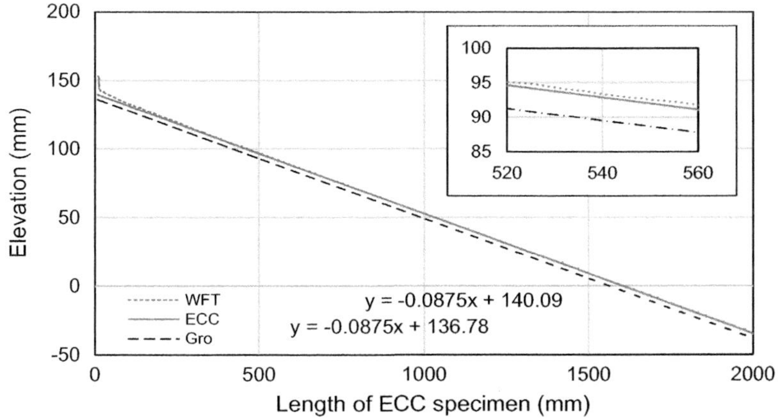

Fig. 12.9 Elevation of water and ECC-Cor Gro based on isoclip method with CFX

Table 12.3 WFT results comparison between lab tests and CFX model

Mixture	MTD	Grade (%)	Water film thickness (mm)			Difference (%)	
			CFX	Lab test	Eq. 10.2	CFX versus Lab (%)	CFX versus Eq. 10.2
ECC-Cor	0.36	1	1.26	1.33	1.30	5.6	3.4
		5	1.00	1.03	1.00	3.3	0.5
ECC-Cor Gro	0.36[a]	1	0.88	0.89	0.85	1.1	3.5
		5	0.53	0.50	0.50	5.6	5.6

[a]On the surface only

As can be seen from the results summarized in Table 12.3, a validated 3D CFX model gives a good correlation compared to results from the laboratory tests and results derived from the equation (Eq. 10.2) for water film thickness on ECC-Cor. Maximum and minimum differences are 5.6% and 3.3%, respectively, is observed between results from the CFX model and laboratory test. Meanwhile, the difference in results between the CFX model and the equation is about 3.4% (maximum) and 0.5% (minimum). The validated 3D CFX model also gives a strong correlation with a lab test and equation results. In comparison to lab tests, the 3D CFX model gives 1.1% and 5.6% differences for 1% and 5% grades, respectively. While in comparison to (Eq. 10.2, the 3D CFX model gives 3.5% and 5.6% differences for 1% and 5% grades, respectively. This means the 3D CFX model can be used to study and design a surface water drainage. This model investigates several parameters, including the effect of macrotexture, the effect of grade, and the effect of grooves on water film thickness.

A higher grade of a specimen, which correlates to the grade or the cross slope of pavement design, leads to thinner WFT. On flat ECC-Cor specimen, the increase of the grade from 1 to 5% will reduce the WFT up to 21%. While for grooved ECC-Cor, the WFT of 5% grade specimen is 44% thinner compared to 1% grade specimen.

Grooves on ECC-Cor specimens reduce the WFT on the surface. With the same total area (2 m $\times$ 0.5 m), the WFT on the grooved ECC-Cor surface will be more than 50% decreased in comparison to the ECC-Cor surface. The number is, however, might be varied on the dimension of the grooves (width, length, and thickness). In consequence, thinner WFT will cause a higher safety for motorists/vehicles as it creates smaller risks of hydroplaning effect between tire of vehicles and pavement in wet condition.

The roughness of the material, which is sand grain roughness in the CFX model, negatively affects the water flow velocity. The higher roughness of the material applied, the slower is the water flow in results. This means the surface water drainage model must be designed carefully. When the wall is designed to have roughness, it will reduce the water flow, especially on the wall parts. As a results, the water flow in the middle part will flow faster that the water in the wall area part.

References

1. ANSYS (2019) ANSYS CFX. ANSYS Workbench
2. Ge S, Celikin M, Isac M, Guthrie RIL (2014) Mathematical modeling and microstructure analysis of Al–Mg–Sc–Zr Alloy Strips produced by Horizontal Single Belt Casting (HSBC). ISIJ Int 54(2):294–303. https://doi.org/10.2355/isijinternational.54.294
3. Kays W, Crawford M (1993) Convective heat and mass transfer. McGraw-Hill, Singapore
4. Li VC (2003) On engineered cementitious composites (ECC): a review of the material and its applications. Adv Concrete Technol 1(3):215–230
5. Schlichting H (1955) Boundary-layer theory. McGraw-Hill, New York
6. Seracino R (2004) FRP composites in civil engineering. CICE 2004, Adelaide, Australia, 8–10 Dec. A.A. Balkema, London

Chapter 13
Study of Correlation Between Skid Resistance and Surface Water Drainage Performance

13.1 Introduction

This chapter elaborates on the correlation between the skid resistance and the surface water drainage performance-based on analysis from data collected and numerical simulation modelled. It has been discussed in previous chapters that the safety of the riding experience is mainly related to skid resistance and surface water drainage. A previous study mentioned a possible case where one method can provide good skid resistance but not surface water drainage [5]. This fact emphasizes that those two criteria should be designed well that they are contributing to the safety comprehensively.

In this study, corundum as fine aggregate is first incorporated into the ECC mixture (ECC-Cor) to improve the surface microtexture. The resulting ECC-Cor possesses satisfactory mechanical performance and high skid resistance (BPN > 70). ECC-Cor also shows good durability as the microtexture and macrotexture can last for more than 100,000 load cycles. The addition of corundum aggregates significantly enhances the microtexture and skid resistance. An improvement on macrotexture and surface drainage is marginal. Further surface treatment with grooves successfully improves the functional performance, not only by minimizing the risk of hydroplaning (water outflow time <4 s) but also by improving the skid resistance (BPN > 90). An empirical model, which correlates water drainage to macrotexture and pavement geometries of ECC, is proposed.

13.2 Experimental Program

13.2.1 Materials and Mixture Design

Type I Portland cement, class F fly ash (FA), micro-silica sand (mean and maximum grain size of 110 μm and 200 μm, respectively), polyvinyl alcohol (PVA) fibers

A. A. Bawono, *Engineered Cementitious Composites for Electrified Roadway in Megacities*, Springer Theses, https://doi.org/10.1007/978-3-030-88542-7_13

(length of 12 mm, a diameter of 39 μm, a nominal tensile strength of 1600 MPa), high-range water-reducing (HRWR) were used to prepare the mix.

In this study, corundum with a maximum particle size of 2.38 mm was used as fine aggregates for improving the surface textures of ECC. The material size distribution of corundum is given in Fig. 13.1. Corundum is a crystalline form of aluminum oxide (Al2O3) with extremely high hardness [7]. It is rated at nine on the Mohs scale of mineral hardness, where diamond rated as 10, whereas the most common mineral in aggregates, quartz, only rated as 7 [1, 12]. Based on field experience in Swiss, corundum was used to ensure high skid resistance for the surface of concrete pavement [7].

Table 13.1 summarizes the mixture design of ECC incorporating a different amount of corundum (ECC-Cor). A control ECC mixture without the addition of corundum (ECC-Cor 0) was also prepared. All ECC-Cor mixtures have a low water-to-cementitious material ratio (w/cm) of 0.27. Micro-silica sand was replaced with corundum at different dosage, ranging from 50 to 100%, and the mixture was

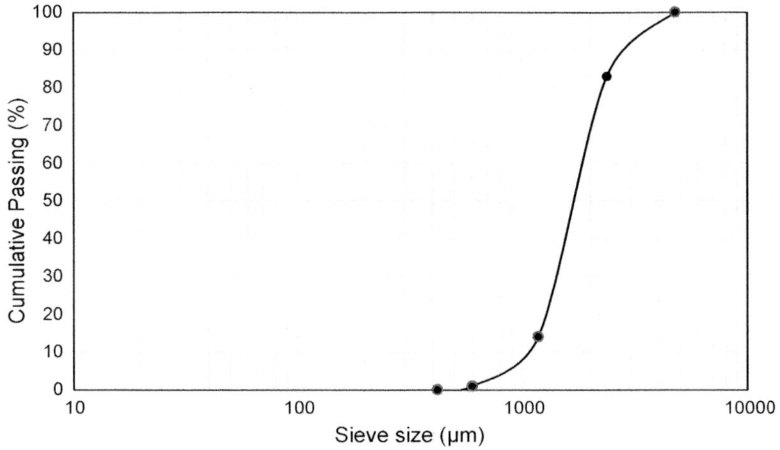

Fig. 13.1 Size distribution of corundum aggregate

Table 13.1 Mixture design of ECC-Cor (kg/m^3)

Mixture	Cement	Fly ash	Sand	Corundum	Water	PVA	HRWRA
ECC-Cor 0	559	671	447	0	327	26	2.3
ECC-Cor 50	559	671	224	224	327	26	2.3
ECC-Cor 60	559	671	179	268	327	26	2.3
ECC-Cor 70	559	671	134	313	327	26	2.3
ECC-Cor 80	559	671	89	358	327	26	2.3
ECC-Cor 90	559	671	45	402	327	26	2.3
ECC-Cor 100	559	671	0	447	327	26	2.3

denoted as 'ECC-Cor #', indicating # with % of micro-silica sand was replaced with corundum.

13.2.2 Specimen Preparation

ECC-Cor mixture was prepared in a planetary mixer. Cement, fly ash, and micro-silica sand are dried mixed first. Water was then added, followed by the addition of superplasticizer and mixed for 3–5 min until the mix achieved the required consistency. Lastly, the PVA fiber and corundum were added and mixed for another 3–5 min. The fresh mixture was cast into a prism (280 mm by 70 mm by 50 mm) and plate (500 mm by 300 mm by 10 mm) moulds catered for various types of tests. Specimens were removed from the mould after one day and were cured at normal laboratory air (25 ± 3 °C, 70%RH) for another 27 days before further testing. At the pre-determined age, either one day or three days, the top surface of plate specimens was treated with a steel brush to expose the corundum on the surface.

An alternative specimen preparation method was also investigated in the current study. The alternative method has the objective to have a sufficient skid resistance with less corundum used. Instead of mixing corundum into the ECC mixture, corundum (150 kg/m^3) was first spread and distributed uniformly on the base of the plate moulds, followed by casting ECC-Cor 0 into the mould. The specimen is named ECC-Cor 0-S. Plate specimens were removed from the mould after one day, followed by curing at normal laboratory air (25 ± 3 °C, 70%RH) for another 27 days before further testing.

Parallel grooves were created on surface of ECC-Cor 80 specimens (500 mm by 300 mm by 10 mm) using a concrete cutter machine to study the effects of surface macrotexture on skid resistance and surface water drainage performance of ECC-Cor. The concrete cutter was equipped with a blade of 350 mm in diameter and 3.2 mm in thickness. The blade position is adjustable, so it can slide forward and backwards, while the level can be fixed to make a groove with the specified depth. Four macrotextures were evaluated, and the corresponding groove dimensions are summarized in Table 37.

13.2.3 Test Program

Table 13.2 summarizes the test program of the current study. Mechanical properties, skid resistance, surface water drainage performance, and durability were evaluated using the four-point bending test (FPBT), the British Pendulum test (BPT) and the sand patch method (SPM), the outflow meter test, and the wheel tracking test (WTT), respectively.

Microscopic skid resistance was evaluated following ASTM Standard Test Method for Measuring Surface Frictional Properties Using the British Pendulum

Table 13.2 Test program

Mixture	Mechanical properties		Skid resistance		Surface water drainage performance	Durability
	MOR	Ductility	Microtexture	Macrotexture		
	FPBT	Inverse	BPT	SPM	OFT	WTT
ECC-Cor 0	Y	Y	Y	Y		
ECC-Cor 50	Y	Y	Y	Y		Y
ECC-Cor 60	Y	Y	Y	Y		
ECC-Cor 70	Y	Y	Y	Y		
ECC-Cor 80	Y	Y	Y	Y	Y	
ECC-Cor 80-Gro 1			Y	Y	Y	
ECC-Cor 80-Gro 2			Y	Y	Y	
ECC-Cor 80-Gro 3			Y	Y	Y	
ECC-Cor 80-Gro 4			Y	Y	Y	
ECC-Cor 90	Y	Y	Y	Y		
ECC-Cor 100	Y	Y	Y	Y		Y
ECC-Cor 0-S			Y	Y		

Tester [3]. The test results are reported as British Pendulum Number (BPN). Plate specimens for BPT have a dimension of 500 mm by 300 mm by 10 mm. The tests were conducted in wet test conditions by spraying water on the specimens. At least three specimens were evaluated for each mix, and a minimum of five tests were performed for each specimen. BPT was also used to demonstrate the effects of macrotexture provided by the grooves. Previous studies mentioned that BPT is reliable for defining both microtexture and macrotexture [9, 10]. However, the macroscopic skid resistance was mainly evaluated using SPM, a volumetric-based spot test method with a sand patch. The method is based on ASTM Standard [4, 11] and can be used to determine the Mean Texture Depth (MTD) of pavement, which reflects the macrotexture characteristic of the pavement surface.

Surface water drainage performance was evaluated using the outflow meter (OFM) test in accordance with ASTM E2380M-15, Standard Test Method for Measuring

Pavement Texture Drainage Using an Outflow Meter. The OFM consists of a transparent cylinder that rests on a rubber placed on the surface of the specimen. It measures the duration of a known quantity of water taken to run through voids in the pavement texture (i.e., water outflow time, OFT). It intends to measure the pavement's ability to relieve pressure from the surface of the tires, thus indicating hydroplaning potential under wet conditions. A faster run time indicates a thinner film of water may exist between the tire and pavement surface [2]. Surface water drainage performance of ECC-Cor 80 specimens with different groove dimensions was evaluated, and ECC-Cor 0 was also tested as the control. The cross slope and the grade of the specimens were confirmed to be on a zero-grade using a digital inclinometer to avoid the external effect. A minimum of three specimens were prepared, with a minimum of five tests were performed for each specimen. The test used water with a temperature of 28 °C in an outdoor environment with an air temperature of 30 °C.

Effects of geometrical road design with different cross slope and grade to the surface water drainage performance were studied. A tool with two layers of a wooden plate was made to evaluate the effects of geometrical design on surface water drainage, as shown in Fig. 13.2. The purpose of the top plate is to simulate the grade of the pavement. The top plate can be adjusted from 0.0 to 4.0% grade using hex bolts. The bottom plate, which can be adjusted from 0.0 to 5.0%, was designed to evaluate the effect of the cross slope on the surface water drainage. The concrete specimens were placed on top of the wooden plate tool when the OFTs were conducted to check the surface drainage with different cross slope. The cross slope was checked using an inclinometer in units of mm/m with a 0.10 mm/m error. It is important to measure the inclination by putting the inclinometer above the specimens rather than above the wooden plates, as the wooden plates might be bent due to the weight of the specimen. Afterwards, measurements of the water outflow time with different slopes were conducted. Minimum of three specimens were prepared, with a minimum of five tests were performed for each specimen. The test was done using

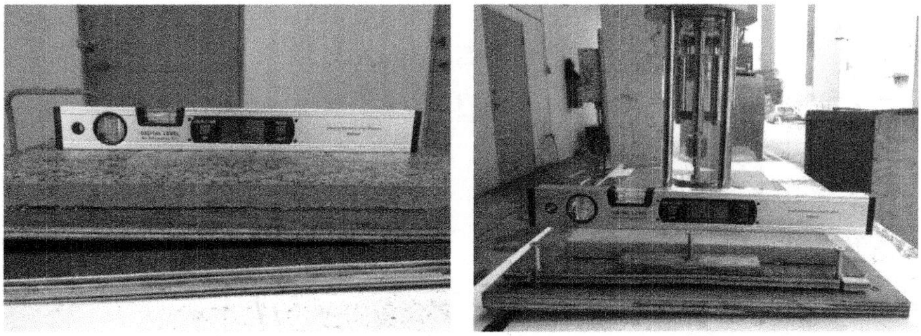

(a) (b)

Fig. 13.2 Adjustment mechanism for **a** grade and **b** cross slope of the specimen water with a temperature of 28 °C in an outdoor environment with an air temperature of 30 °C.

13.3 Results

13.3.1 Skid Resistance

BPN results of ECC-Cor specimens are presented in Fig. 13.3. As can be seen, without the addition of corundum, ECC-Cor 0 has a BPN of 53 ± 9.1, which is lower than the required skid resistance of 55 for new roadways in accordance with the local standard set by the Land Transport Authority of Singapore. This suggests that ECC-Cor 0 cannot be used for road application unless surface treatments are introduced.

ECC mixed with corundum (i.e., ECC-Cor 50 to 100) with surface exposure after three days shows some improvement in skid resistance as most of the results are above 55 BPN on average, except for ECC-Cor 90. The moderate improvement may be attributed to the delayed surface exposure process at day 3, where the matrix, a fiber-reinforced paste, was very hard to be removed and thus, fewer corundum aggregates were exposed on the surface. It can be seen from the same figure that there is no clear correlation between the amount of corundum aggregates and skid resistance of ECC-Cor. This again may be attributed to the delayed surface exposure process at day 3, which is not able to expose the surface corundum evenly, as evident by the large variation of the BPN results.

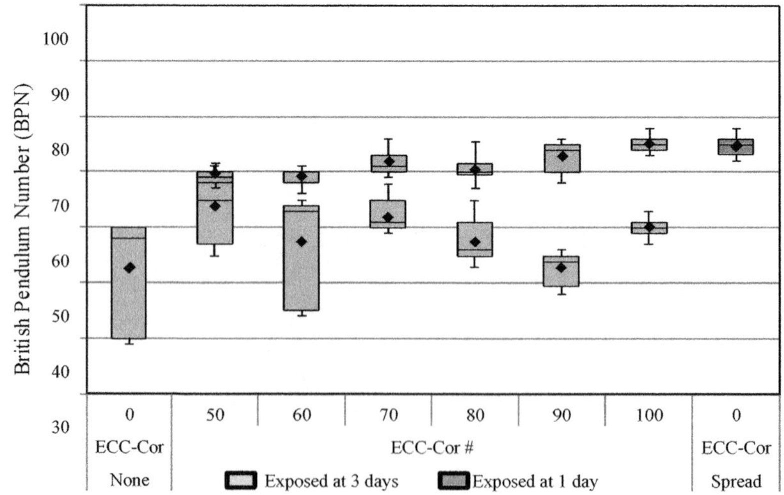

Fig. 13.3 BPN of ECC-Cor

A significant improvement of skid resistance of ECC-Cor with exposing process one day after casting was observed (BPN above 65). From Fig. 13.3, it can be observed that skid resistance (BPN) increases with an increasing amount of corundum used in the ECC-Cor mix. The ECC-Cor 100 possesses a BPN of 75, about 40% improvement compared to ECC-Cor 0. This is because the surface exposure process was carried out on day one, and the paste can be removed easily and evenly, as evident by the reduced variation of the BPN results. High skid resistance is also observed for the ECC-Cor 0-S specimen (BPN of around 75, comparable to the ECC-Cor 100). Even with much less corundum was included in the ECC-Cor 0-S specimen (150 kg/m^3) as compared to other ECC-Cor mixes (e.g., 225 and 450 kg/m^3 for ECC-Cor 50 and 100 mixes, respectively). Since skid resistance is a surface property, ECC-Cor 0-S represents a potential approach to strategically use corundum to improve the surface property while keeping the superior mechanical performance of the original ECC. Ultimately, this approach may be applied for both cast-in-situ and precast construction.

As shown in Fig. 13.4, the MTD of ECC-Cor 0 is very low, a with surface texture of only 0.09 $\pm$ 0.005 mm. Replacement of 50% corundum (ECC-Cor 50) remarkably increases the MTD to 0.36 $\pm$ 0.02 mm. Further increment of corundum dosage does not lead to continuous enhancement of MTD. On average, the MTD of ECC-Cor # is about 0.35 $\pm$ 0.05 mm, which is three times that of ECC-Cor 0. The variance of MTD results may be due to the fact that many results were drawn from at least three specimens for each mix, and a minimum of five tests were performed for each specimen. Arguably, sand patch test procedures have shown range poor to fair repeatability [13].

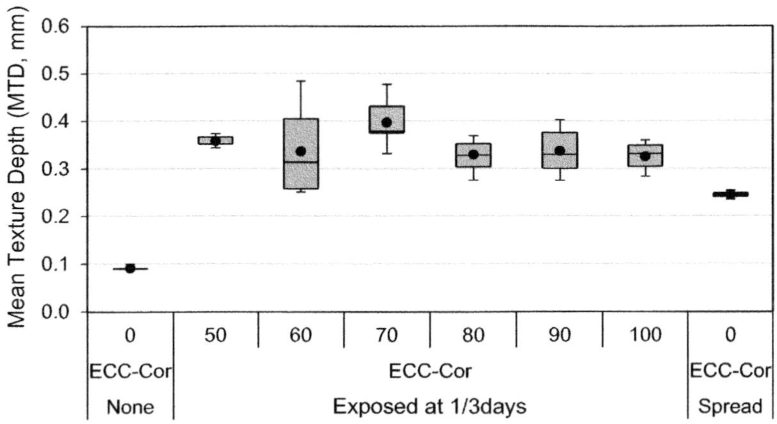

Fig. 13.4 MTD of ECC-Cor

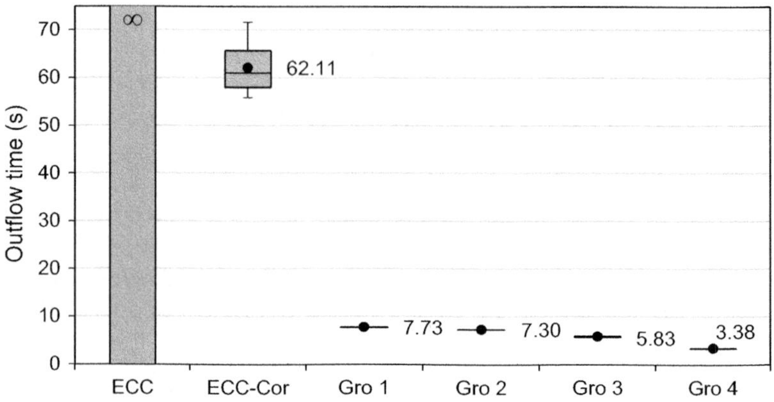

Fig. 13.5 OFT of ECC-Cor

13.3.2 Surface Water Drainage Performance

Surface water drainage performance is governed by the surface macrotexture [8]. A long OFT is indicative of a smooth macrotexture surface, while a short OFT indicates rough macrotexture surface textures. Since ECC-Cor 50 to 100 have similar MTD, only ECC-Cor 80 was measured using OFM to evaluate its surface water drainage performance and compared with ECC-Cor 0. As can be seen in Fig. 13.5, the OFT of ECC-Cor 0 cannot be measured as the surface is very smooth without the inclusion of corundum. The absence of macrotextures creates no gaps between the rubber of the OFM and the surface of the specimen. As a result, the water stays inside the OFM device. The average OFT for ECC-Cor 80 is 62.11 $\pm$ 4.5 s. The macrotextures created by the inclusion of corundum improve the surface water drainage performance to a certain level. Yet, the OFT is still remarkably high that may lead to the risk of hydroplaning effect. In fact, several Department of Transportation for states in the US requires OFT of 9–10 s as the maximum limit [5]. Therefore, further treatments are necessary to fulfil the surface water drainage performance requirement for ECC-Cor 80.

13.3.3 Effects on Grooving Treatment on Surface Water Drainage Performance and Skid Resistance

Table 56 summarizes the OFT, MTD, and BPN of ECC-Cor 80 specimens with different surface grooving treatment (i.e., ECC-Cor 80-Gro 1 to 4) compared with that of ECC-Cor 0 and ECC-Cor 80 as the control. As can be seen, the OFTs for the grooved ECC-Cor 80 specimens are further reduced from 62.11 $\pm$ 4.5 s to 7.73 $\pm$ 0.05 s (ECC-Cor-Gro 1). With reduced land area, the surface water drainage performance further increases. ECC-Cor 80-Gro 4 with a land area of 7.5 ± 0.5 mm shows the shortest OFT of only 3.38 ± 0.08 s. This is because grooving treatment greatly enhances the macrotexture of the specimen, and MTD increases significantly with reducing land area (Fig. 13.6). The combination of the microtexture provided by the corundum and the macrotexture provided by the grooves remarkably enhances the drainage of water through the space between the rubber ring of the OFM device and the specimen surface (Table 13.3).

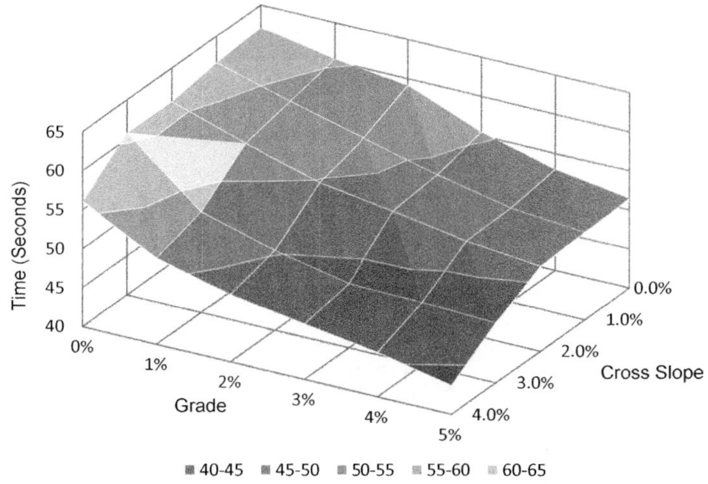

(a)

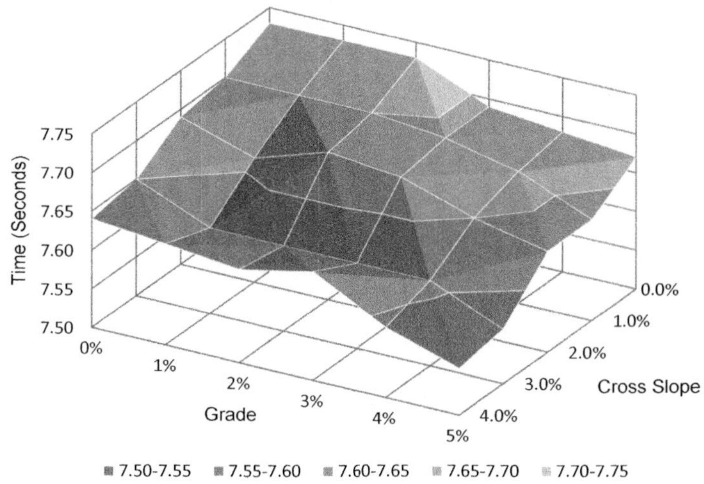

(b)

Fig. 13.6 OFTs for **a** ECC-Cor 80, **b** ECC-Cro 80-Gro 1, **c** ECC-Cro 80-Gro 2, **d** ECC-Cro 80-Gro 3, and **e** ECC-Cro 80-Gro 4 as a function of cross slope and grade

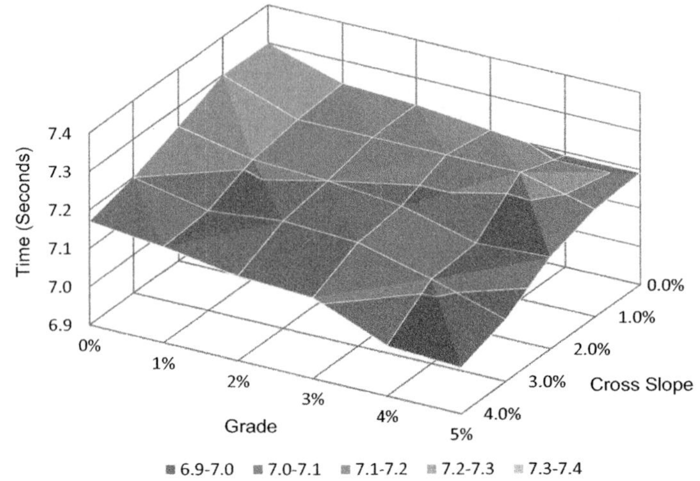

(c)

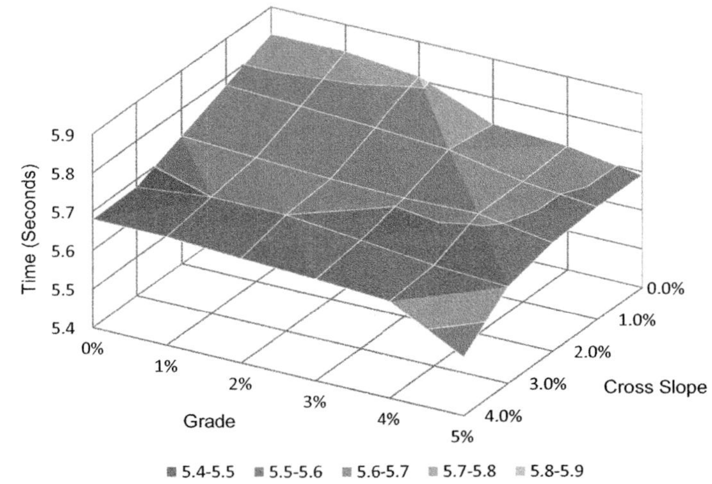

(d)

Fig. 13.6 (continued)

The BPN of the grooved ECC-Cor 80 tested in the transversal direction (i.e., the grooves are perpendicular to the BPT swing direction) is higher than that of the ECC-Cor 80. Furthermore, the BPN increases with reducing the land area. ECC-Cor 80-Gro 4 with a land area of 7.5 ± 0.5 mm shows the highest BPN of 92.80 ± 1.60, which represents a 71% and 30% enhancement as compared to that of ECC-Cor 0 and ECC-Cor 80, respectively. This may be attributed to the increased macrotexture on the surface, which holds the movement of the rubber slider on the pendulum. The

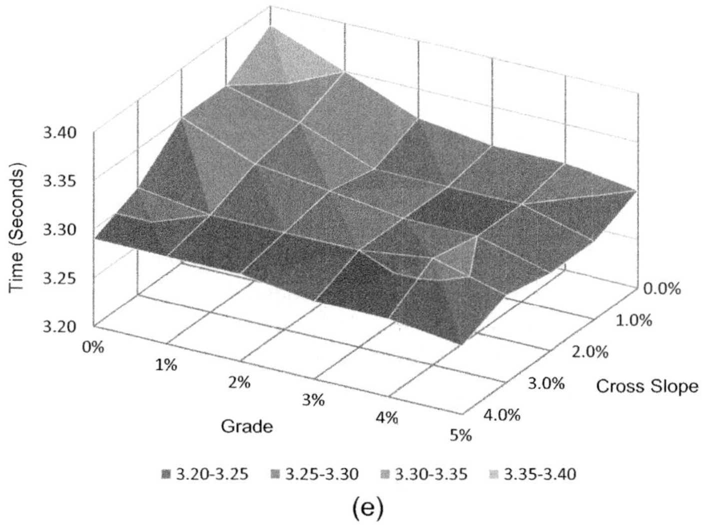

Fig. 13.6 (continued)

Table 13.3 Effect of grooves to water outflow

Specimen	Microtexture—longitudinal (BPN)	Microtexture—transversal (BPN)	Macrotexture, MTD (mm)	Water outflow, OTF (s)
ECC-Cor 0	54.15 ± 2.50		0.09 ± 0.005	>500
ECC-Cor 80	71.43 ± 2.02		0.36 ± 0.019	62.11 ± 4.99
ECC-Cor 80-Gro 1	71.40 ± 1.43	77.50 ± 1.57	0.74 ± 0.030	7.73 ± 0.05
ECC-Cor 80-Gro 2	72.10 ± 0.83	88.60 ± 1.28	0.85 ± 0.030	7.30 ± 0.05
ECC-Cor 80-Gro 3	72.60 ± 0.92	90.50 ± 1.28	1.08 ± 0.054	5.83 ± 0.18
ECC-Cor 80-Gro 4	74.10 ± 0.70	92.80 ± 1.60	1.33 ± 0.059	3.38 ± 0.08

BPN of the grooved ECC-Cor 80 tested in the longitudinal direction (i.e., the grooves are parallel to the BPT swing direction); however, it remains similar to ECC-Cor 80.

13.3.4 Effects of Cross Slope and Grade on Surface Water Drainage Performance

Figure 13.6 plots the OFTs of ECC-Cor 80 with and without grooves at different cross slope and grade, where [6] is the reference to the original paper where the figure appeared. The general trend shows that surface water drainage performance increases with increasing cross slope and grade. This is particularly true for the surface without grooving treatment, i.e., ECC-Cor 80. As shown in Fig. 13.6a, OFT decreases by 9.7% and 17.2%, with a 4% incremental change of cross slope and 5% change of grade, respectively. The OFT decreases by 30% when both cross slope and grade increases by 4% and 5%, respectively. However, it is observed that effects of cross slope and grade are much less pronounced for grooved surface (Fig. 13.7). For example, the OFT of ECC-Cor 80-Gro 1 specimen decreases by 1.2% only when the cross slope increases by 4%, and the OFT decreases by 0.9% only when the grade increases by 5%. Further, OFT decreases by 2.2% when both cross slope and grade increases by 4% and 5%, respectively. This is because grooving treatment has already provided an effective means to drain water and greatly shorten the OFT. Further reduction of OFT by increasing the cross slope and grade is less pronounced. This is particularly true for specimen with larger land area, i.e., ECC-Cor 80-Gro 4.

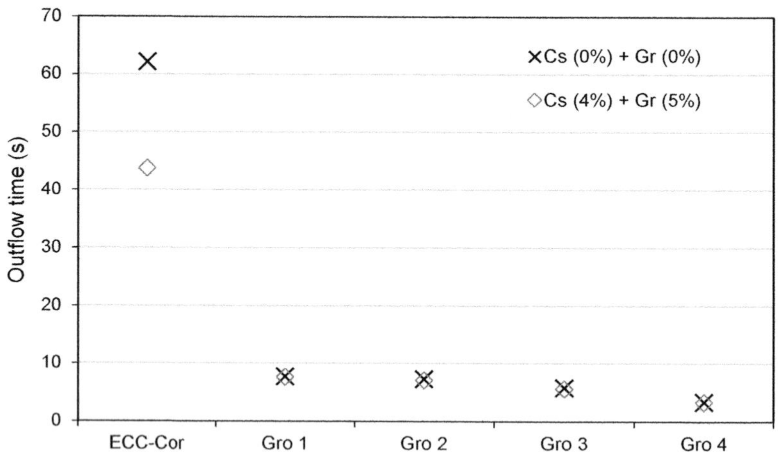

Fig. 13.7 Effects of 4% cross slope and 5% grade on OFT of ECC-Cor 80 with and without grooving treatment

13.3.5 The Correlation Between Skid Resistance and Surface Water Drainage with Effect of Cross Slope and Grade

Based on the experimental results, an empirical formulation associating OFT with MTD and grade and cross slope are proposed by curve fitting.

$$OFT = \frac{5.88MTD^{-2.31}10^{-0.4403MTD^{-1.955}Cs} + 5.88MTD^{-2.31}10^{-0.4403MTD^{-1.955}Gr}}{2} \quad (13.1)$$

where OFT is the outflow time in s; MTD is the Mean Texture Depth in mm; Gr is the grade of pavement in %, and C_S is the cross slope of pavement in %. As shown in Fig. 13.8, the results from Equation (Eq. 13.1) was compared with experimental results as well as the formula proposed in [2] and by Sarsam [13], as can be seen in Eq. 5.7 and Eq. 5.8, respectively.

From the experimental results, it can be seen that, in general, the OFT of ECC reduces with MTD exponentially. Slab with higher cross slope and grade possesses lower OFT at a given MTD. This is particularly true for a slab with low MTD. The current model prediction fits well with the experimental results for both with and without cross slope and grade. The ASTM and Sarsam models, however, deviate significantly with the experimental results, OFT for ECC with low MTD in particular. At higher MTD (>1.2 mm), all three models give a good prediction of the OFT. This is because cross slope and grade play an important role and govern water drainage at low MTD, and the effects of cross slope and grade diminish with increasing MTD. The

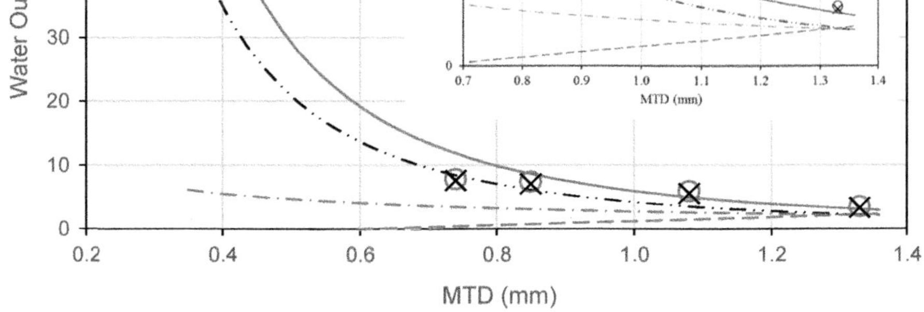

Fig. 13.8 OFT as a function of MTD

ASTM and Sarsam models are developed based on water drainage on conventional concrete and asphalt pavement surfaces which often have higher MTD, and thus effects of cross slope and grade are not considered in these two models. However, due to the absence of coarse aggregates in the ECC mix, low MTD for un-grooved ECC surface is often expected. The proposed model takes cross slope and grade of pavement into account and can predict surface water drainage performance (OFT) for any given MTD and pavement geometries.

References

1. ASTM C78/78M (2018) Test method for flexural strength of concrete (Using Simple Beam with Third-Point Loading). ASTM International, West Conshohocken, PA. https://doi.org/10.1520/C0078_C0078M-18
2. ASTM E2380M-15 (2015) Test method for measuring pavement texture drainage using an outflow meter. ASTM International, West Conshohocken, PA
3. ASTM E303-93 (1993) ASTM standard test method for measuring surface frictional properties using the British Pendulum Tester. ASTM International, West Conshohocken, PA
4. ASTM E965-96 (2006) Standards test method for measuring pavement macrotexture depth using a volumetric technique (reapproved 2006). ASTM International, West Conshohocken, PA
5. Bawono AA, NguyenDinh N, Yang E-H, Lechner B (2017) Evaluation of highway pavement skid resistance performance made of Engineered Cementitious Composite (ECC). In: Tenth international conference on the bearing capacity of roads, railways and airfields, Athens, June. CRC Press, Greece
6. Bawono AA, Lechner B, Yang E-H (2019) Skid resistance and surface water drainage performance of engineered cementitious composites for pavement applications. Cement Concr Compos 104:103387. https://doi.org/10.1016/j.cemconcomp.2019.103387
7. Broz ME, Cook RF, Whitney DL (2006) Microhardness, toughness, and modulus of Mohs scale minerals. Am Miner 91(1):135–142
8. Hall J, Smith K, Titus-Glover L, Wambold J, Yager T, Rado Z (2009) Guide for pavement friction, Washington, D.C.
9. Karaca Z (2016) Reliability of BPT on macrotextured surfaces. Int J Innov Sci Res 5(1):611–616
10. Lee K (2004) Analyzing laboratory skid resistance test using finite element modeling. Doctoral Dissertation, National University of Singapore
11. NguyenDinh N, Yang E-H, Lechner B (2014) Precast electrified roadway pavement systems using ECC. In: 12th international symposium on concrete roads, Prague, Czech Republic
12. Qian S, Li VC (2007) Simplified inverse method for determining the tensile strain capacity of strain hardening cementitious composites. J Adv Concr Technol 5(2):235–246
13. Sarsam SI (2012) Field evaluation of asphalt concrete pavement surface texture and skid characteristics. In: 5th Eurasphalt & Eurobitume congress, Istanbul, 13–15th June 2012

Chapter 14
Study of Noise

14.1 Introduction

This chapter presents the study of noise on the pavement made of ECC. In pavement material design, there are three main material properties which are affecting noise, namely density, porosity, and texture. Thus, it was expected that ECC material might produce higher noise due to its high density and less porous properties. However, there is a need to study the noise caused by pavement made of ECC as a reference for future usage as well as identification for an improvement. Corundum is incorporated as fine aggregates into the ECC mixture (ECC-Cor) to improve the surface microtexture. The ECC-Cor results in satisfactory performance of high skid resistance (BPN > 70). Further surface treatment with grooves successfully improves the functional performance, not only by minimizing the risk of hydroplaning (water outflow time < 4 s) but also by improving the skid resistance (BPN > 90). The noise that comes from the same texture on ECC-Cor is further investigated. Singapore has transformed into a highly developed city, with most of the population live in urban areas. Due to a limited area, the most residential area is developed close to expressways and major arterial roads with high traffic densities which cause noise pollution. Strait times [8] reported that communities have public complaints about noise pollution. Quieter pavements were considered as one of the solutions that authorities should provide. Yet, no studies have been reported on noise level for pavement made of ECC.

The contact-free measuring method applied to the specimens allows for a non-destructive determination of the surface roughness (macrotexture) of slab surfaces. The noise level is then simulated and predicted by using SPERoN software based on the approach that pre-processed input data is used [9].

A. A. Bawono, *Engineered Cementitious Composites for Electrified Roadway in Megacities*, Springer Theses, https://doi.org/10.1007/978-3-030-88542-7_14

14.2 Experimental Program

14.2.1 Materials and Mixture Design

The mixture design used in this study is mixed ECC with corundum (ECC-Cor), as has been discussed in Chap. 6. ECC-Cor is using a Type I Portland cement, class F fly ash (FA), micro-silica sand (mean and maximum grain size of 110 μm and 200 μm, respectively), polyvinyl alcohol (PVA) fibers (length of 12 mm, a diameter of 39 μm, a nominal tensile strength of 1600 MPa), high-range water-reducing (HRWR) were used to preparing the mix.

Corundum with a maximum particle size of 2.38 mm was used as fine aggregates for improving the surface textures of ECC. The material size distribution of corundum is given in Fig. 13.1. A control ECC mixture without the addition of corundum (ECC-Cor 0) was also prepared. All ECC-Cor mixtures have a low water-to-cementitious material ratio (w/cm) of 0.27. Micro-silica sand was replaced with corundum at 80% dosage.

14.2.2 Specimen Preparation

ECC-Cor mixture was prepared in a planetary mixer. Cement, fly ash, and micro-silica sand are dried mixed first. Water was then added, followed by the addition of superplasticizer and mixed for 3–5 min until the mix achieved the required consistency. Lastly, the PVA fiber and corundum were added and mixed for another 3–5 min. The fresh mixture was cast into steel formwork with a dimension of 2000 mm by 500 mm by 80 mm. Specimens were removed from the mould after one day and were cured at normal laboratory air (25 ± 3 °C, 70%RH) for another 27 days before further testing. At the pre-determined age, one day, the top surface of plate specimens was treated with a steel brush to expose the corundum on the surface.

Parallel grooves were created on the surface of ECC-Cor 80 specimens by using a silicone rubber to study the effects of surface macrotexture on skid resistance and surface water drainage performance of ECC-Cor. The design used ECC-Cor Gro 01 is shown in Table 7.2. Three slab specimens were prepared: ECC, ECC-Cor 80, and ECC-Cor 80-Gro 1.

14.2.3 Test Program

This study consists of three stages: laboratory measurement, numerical study with finite element analysis (FEA), and noise prediction analysis. The firstly is to determine the actual tire contact stresses and contact area using the Tekscan I-Scan measuring device. The output will be used for the input to FEA to comprehensively study pavement response. Finite element (FE) tools ANSYS will be employed in this study. The results, however, are already discussed in the chapter of the study of mechanical properties, showing there is no significant effect of grooves on the ECC slab.

Secondly, the contact-free measuring method applied allows for a non-destructive determination of the surface roughness (macrotexture) of slab surfaces. The measurement and description of the surface texture is carried out by using a profilometer, which set according to ISO 13473 "Characterization of pavement texture by use of surface profiles", Parts 1–4 [7]. The slab surface is scanned using a laser triangulation measuring system over the whole length according to the method of continuous measurements. The profilometer is connected to the power source and a computer that has the controls and software to record and save the measurements done by the laser. The mean profile depth of the slab is measured by the equipment based on a triangulation laser measurement system. The percentage of the laser beams which fail to reflect is called the dropout rate. This determines the accuracy of the measurement. According to ISO 18743, a dropout percentage of 5% is allowed for one trial. The texture of three slabs: ECC, ECC-Cor, and ECC-Cor Gro 1 were measured with the surface profilometer, which can be seen in Fig. 14.1, to evaluating the tyre-road-noise. The results, which is measured as texture depth, is required as an input for the software. The measured data must contain at least six parallel measurement lines with a length of at least 2 m each. Therefore, the tested ECC-Cor slabs were made with a minimum length of 2 m.

As shown in Fig. 14.1, the laser scanning equipment is aligned parallelly over the slab. Slopes or slants in both the vertical and horizontal directions are avoided to allow the movement of the laser scanner in a straight line. Once the setup is done, a trial run is conducted over the full length of the slab to ensure the laser runs on a straight line. A total of 96-line areas are scanned along the length of the slab with

Fig. 14.1 Surface profilometer measuring the texture of ECC slabs

an offset of 2 mm. The total run time for 96-line areas on one slab is approximately 20 min. The process is repeated for all three slabs.

The resolution of the applied measuring system is 200 μm horizontally and 8 μm vertically. Data will be then processed to predict the noise level by evaluating the MPD-value (according to ISO 13473-1) and the shape factor of the profile; these two values give a good indication of whether the road surface is suitable to reduce the tire-road noises.

$$p^2{}_{coast-by} = p^2{}_{vib} + p^2{}_{air} + p^2{}_{cav} + p^2{}_{res} \tag{14.1}$$

$$p^2{}_{vib} = aF_c^2\Gamma^{\alpha 1}B^{\alpha 2}S^{\alpha 3} \tag{14.2}$$

$$p^2{}_{air} = b\left(F_c^2\Gamma^{-1.5}S^{-2}\right)^{\beta 1}B^{\alpha 2}V^4 \tag{14.3}$$

$$p^2{}_{cav} = cG_{pattern}^{y1} \tag{14.4}$$

$$p^2{}_{res} = dV^{\delta 1} \tag{14.5}$$

The tyre-road noise is predicted by using SPERoN (Statistical Physical Explanation of Rolling Noise) software, which is based on the pre-processed input data of texture [9]. The SPERoN model calculates the coast-by noise of a vehicle by considering four parameters tyre vibrations, the aerodynamic sound sources, the tire cavity modes, and the residual sound sources [9], which then followed with the statistical model [2]. The equation can be seen below.

Where: a, b, c, d are regression coefficients per 3rd octave band; $\alpha_1, \alpha_2, \alpha_3$, $b_{1,}$, γ_1, δ_1 are exponents; F_c is contact force in N; Γ is tyre/road contact air flow resistance in Pa s/m; B tyre width in m; S is tyre tread stiffness in N/m; $G_{pattern}$ is spectral power of the tread pattern variation; V is spectral power of the tread pattern variation driving speed in m/s.

Thus, the input characteristics of the road surface are necessary to obtain results on the coast-by noise, including road texture (roughness of the road surface), acoustical impedance (a measure for the sound reflection and absorption properties of a road surface with 0% as totally reflecting and 100% as totally absorbing), airflow resistance, and mechanical impedance (as the ratio of the input force and vibration velocity measured at a certain point of the structure. The SPERoN calculations were performed with a tyre called "Michelin Energy Saver", which used as a standard tyre, with a speed of 50 km/h [4]. The SPERoN model has been validated by comparing eight coast-by measurement results. The results show that the total sound levels for all of the investigated road surfaces have a mean precision of 1.5 dB [1, 3].

14.3 Results

14.3.1 Texture Depth Measurement with Profilometer

Surface texture depth from three slabs: ECC, ECC-Cor, and ECC-Cor Gro have been measured. As can be seen Fig. 14.2, the texture depth of ECC is flatter in comparison to the other surfaces with only less than 100 μm deviation. Meanwhile, ECC-Cor has a rougher texture depth with about 350 μm deviation as it has a rougher texture because of the corundum. Lastly, ECC-Cor Gro has a similar rough texture with ECC-Cor with additional spikes due to the grooves. The texture depth results from the laser profilometer are in line with results drawn from the sand patch test where ECC-Cor has more macrotexture than ECC, and ECC-Cor Gro has more texture than ECC-Cor.

14.3.2 Macrotexture (MPD) and Shape Factor

Throughout the measurement with the laser profilometer, two parameters were measured, including macrotexture, in mean profile depth (MPD), and the shape factor for each slab. These two parameters are necessary as then later used for the calculation of the total sound level. Previous studies mentioned that two road surfaces with identical texture amplitude spectra do not necessarily show identical acoustical behavior. Therefore, the shape factor needs to be considered as it corresponds to the percentage contact length within the characteristics profile length [2, 6] (Table 14.1 and Fig. 14.3).

14.3.3 Sound Level

The calculation with SPERoN was done for three slabs based on texture depth as one of the main inputs [4]. There are four different noise levels produced with different parameters, including vibration, air, cavity, and residual noise. Those four noise parameters have been explained in (Eq. 36). The results of sound pressure level, which are summarized from Tables 14.2, 14.3, 14.4 and 14.5, are explained based on the octave band center frequency, reported in Hertz (Hz). Hertz in this noise known as the frequency of a sound as how fast the small air pressure changes are occurring. Similar to sound level, the perception of frequency is also non-linear.

Two type of sound level are considered as the main sources, including vibration and air [5]. As shown in Fig. 14.2, those sound levels for ECC-Cor is higher than ECC. The higher sound level of ECC-Cor occurred because of the lower shape factor (G) due to a rougher surface as results of the existence of corundum aggregate on the

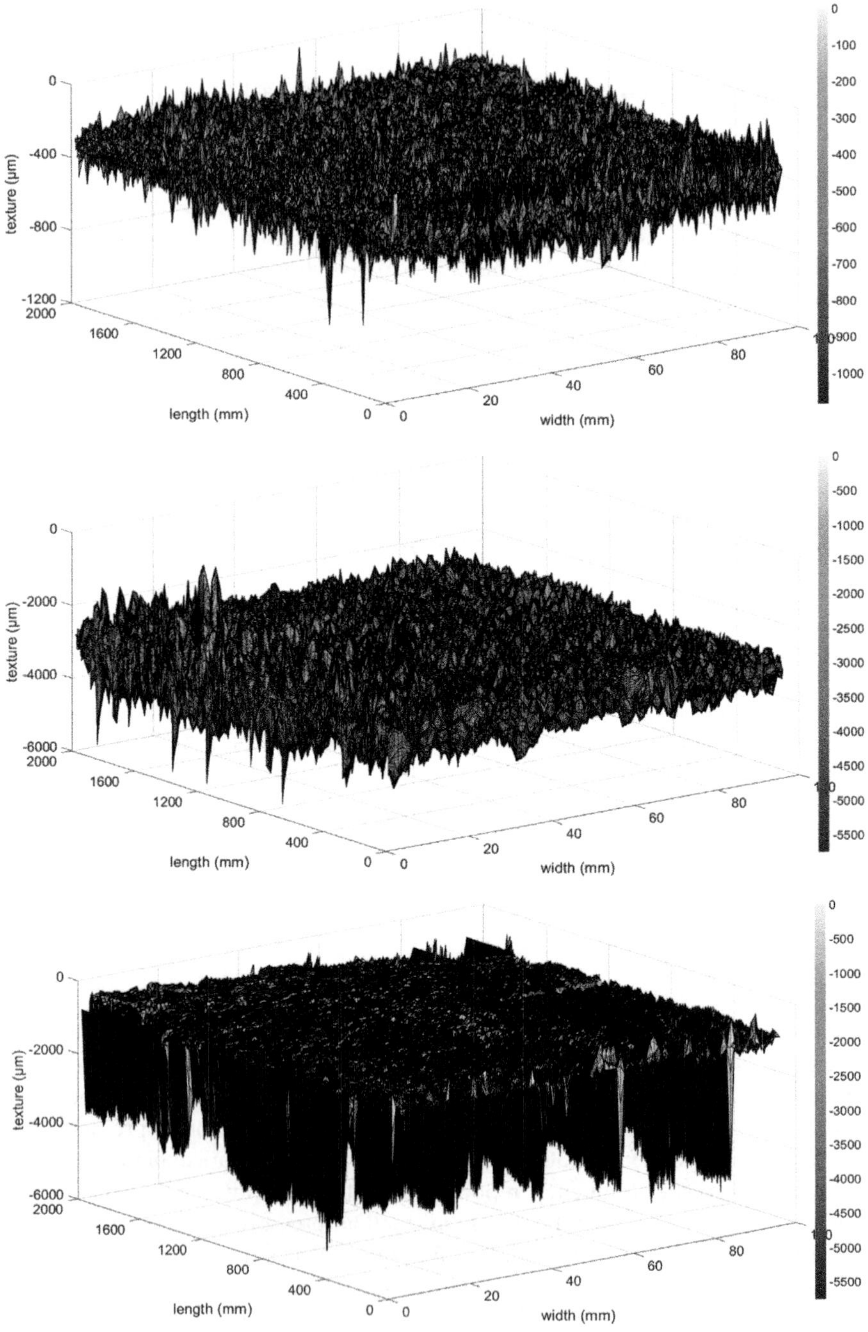

Fig. 14.2 Texture depth of **a** ECC, **b** ECC-Cor, and **c** ECC-Cor Gro

Table 14.1 MPD and shape factor of ECC, ECC-Cor, and ECC-Cor-Gro 1

Texture	MPD	G
	(mm)	(%)
ECC	0.10 ± 0.03	69.8 ± 0.19
ECC-Cor 80	0.61 ± 0.22	55.4 ± 0.17
ECC-Cor 80 Gro 1	0.21 ± 0.20	59.1 ± 0.20

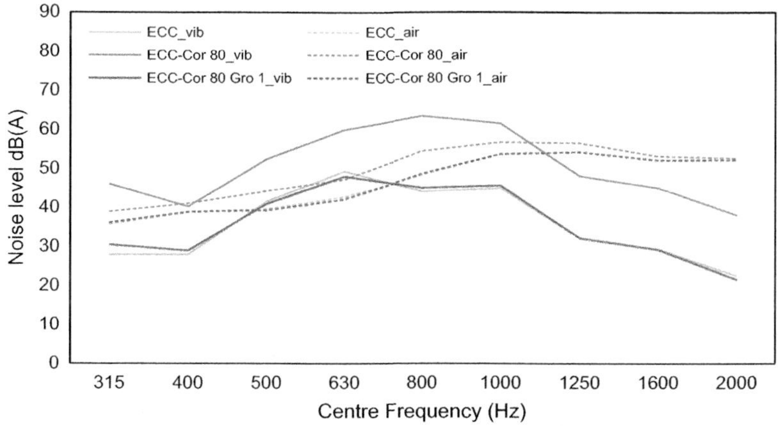

Fig. 14.3 Noise level of ECC, ECC-Cor, and ECC-Cor-Gro 1 based on vibration and air

Table 14.2 Sound level (vibration) of ECC, ECC-Cor, and ECC-Cor-Gro 1

Texture	L vibration in dB (A)—based on the octave band center frequency (Hz)									
	315	400	500	630	800	1000	1250	1600	2000	Tot
ECC	27.9	27.9	41.4	49.2	44.0	44.8	32.0	29.2	22.4	51.9
ECC-Cor 80	45.9	40.1	52.3	59.8	63.3	61.4	47.9	44.9	37.9	66.8
ECC-Cor 80-Gro 1	30.4	28.9	40.8	47.9	44.9	45.5	32.0	29.1	21.4	51.6

Table 14.3 Sound level (air) of ECC, ECC-Cor, and ECC-Cor-Gro 1

Texture	L air in dB (A)—based on the octave band center frequency (Hz)									
	315	400	500	630	800	1000	1250	1600	2000	Tot
ECC	35.6	38.7	39.5	42.6	48.2	53.5	54.1	52.0	52.2	59.6
ECC-Cor 80	38.8	40.9	44.1	47.1	54.3	56.6	56.4	53.0	52.5	62.1
ECC-Cor 80-Gro 1	36.0	38.8	39.2	42.0	48.5	53.6	54.1	52.0	52.1	59.6

ECC surface. The shape factor and the MPD for each design mixture are summarized in Table 57.

Table 14.4 Sound level (cavity) of ECC, ECC-Cor, and ECC-Cor-Gro 1

Texture	L cavity in dB(A)—based on the octave band center frequency (Hz)									
	315	400	500	630	800	1000	1250	1600	2000	Tot
ECC	39.9	44.4	55.4	50.8	56.0	56.2	56.0	19.2	12.4	62.3
ECC-Cor 80	39.9	44.4	55.4	50.8	56.0	56.2	56.0	34.9	27.9	62.4
ECC-Cor 80-Gro 1	39.9	44.4	55.4	50.8	56.0	56.2	56.0	19.1	11.4	62.3

Table 14.5 Sound level (residual) of ECC, ECC-Cor, and ECC-Cor-Gro 1

Texture	L cavity in dB(A)—based on the octave band center frequency (Hz)									
	315	400	500	630	800	1000	1250	1600	2000	Tot
ECC	37.9	44.7	42.9	42.1	38.8	38.6	37.6	35.3	35.6	50.0
ECC-Cor 80	37.9	44.7	42.9	42.1	38.8	38.6	37.6	35.3	35.6	50.0
ECC-Cor 80-Gro 1	37.9	44.7	42.9	42.1	38.8	38.6	37.6	35.3	35.6	50.0

Throughout the measurement with the laser profilometer, two parameters were measured including, macrotexture, in mean profile depth (MPD), and the shape factor for each slab. These two parameters are necessary as then later used for the calculation of the total sound level. Previous studies mentioned that two road surfaces with identical texture amplitude spectra do not necessarily show identical acoustical behavior. Therefore, the shape factor needs to be considered as it corresponds to the percentage contact length within the characteristics profile length.

The other parameters, tyre cavity, is mainly determined by the tyre pattern and less affected by the road surface. Due to the use of same standard tyre, there are no notable differences. The residual component is the cumulation of all airflow noises caused by the frame of the vehicles, which is statistical and only dependent on the velocity. Therefore, all results of the noise level for the slabs are the same.

As results, as can be seen in Table 62, the total sound level of ECC-Cor is higher than ECC, with 69.2 compared to 64.6 dB for ECC-Cor and ECC, respectively. The modification of ECC-Cor by introducing grooves (ECC-Cor-Gro 1) has been successfully reducing the sound level coming from the vibration. The sound level from the air parameter is slightly reduced, mainly in the frequency from 315 to 1250 Hz. The shape factor is then increased for ECC-Cor-Gro 1. As results, the total sound level for ECC-Cor-Gro 1 is reduced significantly from 69.2 (ECC-Cor without grooves) to 64.6 dB, which is lower than the NEA limits of 67 dB for new roads and expressways.

In summary, some findings from the study of noise on pavement made of ECC can be drawn. Those findings including:

- The evaluation of the MPD-value (acc. to ISO 13473–1 (ISO 13473–1 2019)) and the shape factor of the profile; are giving a good indication of whether the road surface is suitable to reduce the tire-road noises.

- The noise level coming from the pavement made of ECC can be up to 65 dB (A), which is quite safe as it is below many standard values regulated by the authorities. However, from the perspective of skid resistance and surface water drainage performance, the ECC slab is not safe to be used for roadways.
- The noise level for ECC-Cor-Gro 1 increased up to 69.2 db(A) because of the lower shape factor (G) due to a rougher surface as results from the existence of corundum aggregate on the ECC surface. The shape factor at ECC-Cor is at 55%, which is very low for dense surfaces.
- It can be seen from the analysis of the different formation mechanism that the biggest differences can be found in the vibrational noises, meaning that ECC-Cor has a rough surface that stimulates the tire to vibrate more. The aeroacoustics parts are as well increased compared to the ECC slab.
- The modification of ECC-Cor by introducing grooves (ECC-Cor-Gro 1) has been successfully reducing the sound level coming from the vibration. The sound level from the air parameter is slightly reduced, mainly in the frequency from 315 to 1250 Hz. The shape factor is then increased for ECC-Cor-Gro 1.
- The ECC-Cor with grooves can be a solution for the (electrified or normal) roadways with pavement made of ECC in Singapore condition, where the NEA regulates the noise limit for 67 dB to new roads and expressways.

References

1. Alves S, Maennel M (2016) Application of Speron to the development of low noise road surfaces. EuroRegio2016, Porto, Portugal, June 13–15
2. Beckenbauer T, Kuijpers A (2001) Prediction of pass-by levels depending on road surface parameters by means of a hybrid model. Internoise 2001, The Hague, Netherlands
3. Beckenbauer T, Klein P, Hamet J-F, Kropp W (2008) Tyre/road noise prediction: a comparison between the SPERoN and HyRoNE models—part 1. J Acoust Soc Am 123(5):3388. https://doi.org/10.1121/1.2934048
4. Ertsey-Bayer M, Bawono AA, Janaki T, Lechner B (2020) Prediction sound level for pavement made of ECC. Müller-BBM, Munich
5. Haider M, Wehr R, Conter M, Kriegisch M, Gasparoni S (2014) Texture and noise characteristics of exposed aggregate concrete road surfaces. Innovative Solutions—Benefiting Society. In: 12th international symposium on concrete roads, Prague, 23 September 2014. European Concrete Paving Association, Prague
6. Hamet J-F, Klein P (2000) Road texture and tire noise. Internoise 2000, Nice, France
7. ISO 13473-1 (2019) Characterization of pavement texture by use of surface profiles
8. Lian GC (2010) Noise pollution in Singapore: when buzz becomes noise. http://wildsingaporenews.blogspot.sg/2010/11/noise-pollution-in-singaporewhen-buzz.html#.WBMRYC196Uk
9. Müller-BBM (2019) SPERoN: rolling noise prediction model. https://www.muellerbbm.com/environment/traffic-environment/speron/

Chapter 15
Study of Self-cleaning and Slip-Proof ECC

15.1 Introduction

This chapter reports the development of a bright and slip-proof ECC with visible light activated photocatalysis property for pavement in tunnels. Different T_iO_2 nano-particles were tested first to evaluate their photocatalysis property under visible light irradiation. The selected T_iO_2 nano-particle was used to develop the bright and slip-proof ECC with the inclusion of white fine aggregates. Effects of fine aggregate content on the mechanical properties and skid resistance of the resulting ECC were investigated. Influences of T_iO_2 dosage on the mechanical properties of ECC were tested and self-cleaning performance and dirt pick-up resistance (DPUR) of ECC under visible light irradiation were studied.

Currently, Singapore is embracing the sustainable development of the land transport to improve the living environment. Emission from the transport sector is one of the main issues in Singapore. Efforts to tackle this issue not only can be done through the improvement of the vehicles wise but also from the pavement infrastructure wise. A self-cleaning pavement can be an excellent alternative solution.

Pavement in tunnels should be designed differently than pavement for other outdoor applications. Visibility is a critical consideration to ensure safety in use and during accidents for pavement in tunnels [19, 29, 31, 37, 48]. The brighter pavement was found to have fewer accidents according to studies done in the UK [20, 27]. By having high surface reflectance of pavement, lower luminance is required, and thus both energy and life cycle cost can be reduced [20, 47]. Another comparison study in the tunnel was done with the conclusion that the power and flux required for the tunnel depend strongly on the reflection properties of the road pavement [44, 51]. By changing the road pavement from asphalt (black) to concrete (grey), it allows 6–10% and 6–8% savings on required flux and power, respectively [51]. Furthermore, concrete is known to have better fire resistance and durability when compared to asphalt. For this reason, more pavements in tunnels are built with concrete nowadays. In fact, some regulations, including the European Directive, state that concrete pavements shall be used for tunnels longer than 1,000 m [29]. Few studies have

A. A. Bawono, *Engineered Cementitious Composites for Electrified Roadway in Megacities*, Springer Theses, https://doi.org/10.1007/978-3-030-88542-7_15

reported the potential use of ECC for rigid pavement applications. Compared to conventional concrete and hot mix asphalt (HMA) overlay systems, ECC overlay offers lower life-cycle cost over a 40-year service life with reduced total life cycle energy by 14%, greenhouse gas (GHG) emissions by 43%, and costs by 40% [55]. Recently, ECC has been applied on an 80 m jointless test track and shows that this material can be produced in large quantity and high quality. The long-term monitoring showed that the material could handle high traffic loads, and the abrasion was low [40]. On top of that, the reflective cracking failure which commonly occurred in an overlay can be eliminated due to the high ductility of ECC material [55].

ECC can be a potential alternative for concrete pavements. However, the critical shortcoming of ECC material, particularly for roadway application, is on the surface texture. As the mix design of ECC omits coarse aggregates and often includes only micro silica sand with particle size less than 300 μm to reduce matrix toughness and to ensure good ductility, it decreases the skid resistance of ECC pavement, which is of great concern for roadway applications with respect to required grip levels [7]. Skid resistance is one of the most important functions in pavement design, as this parameter is responsible for the safety aspects for motorists. Many agree that safety should be prioritized in designing pavement [25, 26]. The skid resistance of pavement is mainly characterized by its micro- and macrotextures. Bawono et al. [8] engaged hard corundum particles into the ECC mix to enhance the skid resistance. However, due to the black color of corundum particles, the surface of the resulting ECC is very dark with low surface reflectivity.

It is desirable to impart additional functions, i.e., self-cleaning and air purification, into concrete for pavement in tunnels. Self-cleaning helps to maintain the brightness and reflectivity of the surface [1, 2, 15], while air purification improves the air quality in tunnels [52]. These two functionalities are often imparted into cement-based materials through engaging ingredients with photo-catalysis properties. Titanium dioxide (T_iO_2) as photo-catalyst was firstly introduced by Fujishima and Honda [21]. Then, in 2003, Cassar [15] did an experimental study in engaging photocatalytic function into concrete by adding T_iO_2 and reported that the resulting concrete is capable to photo-decompose the organic compound sticking on the surface and then to photo-catalyze the transformation of NO_x to NO_3. In 2011, Awadalla et al. [6] concluded that concrete panel incorporating T_iO_2 has the ability to decompose CO_2. At the same time, a study on photocatalytic de-pollution in tunnel irradiated by artificial UV light was reported in Brussels using commercially available TX Active®. UV lamps were used to activate the photocatalytic reaction. However, the field results showed there is no observable reduction of NOx in the tunnel [11, 23]. The current photocatalytically active cement coating, such as TioCem® and TX Active® can effectively remove various pollutants, including volatile organic compounds, by exposing the cement surface to UV light. Zhao et al. [58] incorporated a commercial T_iO_2 into a white ECC mix to engage photo-catalysis property. However, the photo-catalysis property is activated only under UV irradiation. This is because TiO2 used in those studies can only be photoactivated by UV irradiation due to its wide bandgaps (3.0–3.2 eV), which lie in the UV region [22, 35].

15.2 Experimental Program

15.2.1 T_iO_2 Screening, Characterization and Performance Evaluation

Four commercial T_iO_2 were tested in the screening stage: Aeroxide P25 (Evonik, Germany), KRONOClean K7000 (KRONOS, USA), T_iO_2-anatase (Shijiazhuang Kailing, China) and T_iO_2-rutile (Shijiazhuang Kailing, China). Commercially available methylene blue (Sigma-Aldrich, Singapore) and Rhodamine B (Sigma-Aldrich, Singapore) were used for photo-decomposition of dye and self-cleaning experiment, respectively.

The morphology of T_iO_2 was investigated by a field emission scanning electron microscope (FESEM, JSM-7600F). The FESEM combines an electron column with semi-in-lens detectors and an in-the-lens Schottky field emission gun, which is able to deliver ultra-high-resolution secondary electron images with a wide range of probe currents from 1 pA to more than 200 nA. X-ray powder diffraction (XRD) for quantitative phase analysis (QPA) of T_iO_2 was carried out by using a θ/θ diffractometer (XRD Bruker D8 Advance). T_iO_2 samples in round sample holders were placed on the spring plate. The XRD diffractograms were recorded using a Cu Kα radiation at 40 kV and 40 mA with a scan range from 10° to 110° in 2θ at 0.01°/step and exposure time of 0.1 s/step. X-ray photoelectron spectroscopy (XPS) was utilized to characterize the elemental composition of T_iO_2 as well as the chemical state of titanium, oxygen and carbon species. N_2 adsorption–desorption test was utilized to determine the specific surface area of T_iO_2 nanoparticles using Brunauer–Emmett–Teller (BET) theory. The optical properties of T_iO_2 were evaluated by an UV–vis spectrophotometer equipped with a diffuse reflectance accessory (Shimadzu UV-2600). From reflectance spectra, the optical bandgap of commercial T_iO_2 was calculated using the Kubelka–Munk equation. The surface charge of T_iO_2 was assessed by measuring their electro-kinetic potential (zeta potential) via a zeta potential analyzer (Malvern Zetasizer, Nano ZS).

The photo-decomposition of methylene blue was observed based on the absorption spectroscopic technique. In a typical procedure, 50 mg of T_iO_2 was dispersed in 50 ml of methylene blue solution (10 μM) in a 100 mL Pyrex beaker. The mixture was then stirred under the dark condition for 30 min to reach the adsorption equilibrium. After which, the solution was irradiated under visible light for up to 4 h with incessant stirring. A 500 W long arc xenon lamp with a UVIRCUT420 optical filter was used to provide the visible light source. The dye solution was sampled periodically to check the degree of adsorption (in the first 30 min) and photo-decomposition (after the light source is on) of methylene blue. The sample was first centrifuged to separate the T_iO_2 particles from the solution. The light absorption of photo-reacted solution at 660 nm (i.e., the wavelength corresponding to the maximum light absorption of methylene blue) was measured using an UV–visible spectrophotometer (Varian Cary 5000) and compared with that of the pristine methylene blue solution to evaluate the degree of adsorption (in the first 30 min) and decomposition (after the light source is

Table 15.1 Properties and geometry of the PE fibers (Spectra® 1000, Honeywell)

Diameter (μm)	Length (mm)	Tensile strength (MPa)	Elastic modulus (GPa)	Density (g/cm^3)
23	19	3250	113	0.96

on). A control test (i.e., same methylene blue solution without the addition of T_iO_2 underwent the same adsorption and irradiation exposures) was also performed to evaluate potential decomposition of methylene blue due to direct photolysis.

15.2.2 Material

Type I white Portland cement, high strength and high modulus polyethylene (PE) fibers (Spectra® 1000, Honeywell), high-range water reducer (HRWR, ADVA® 181, GCP Applied Technologies), methycellulose-based viscosity modifying agent (VMA), white silica sand, and T_iO_2 nano-particles were used to formulate white ECC. Based on the results of T_iO_2 screening tests in Sect. 2.1, KRONOClean K7000 (KRONOS, USA) was selected to impart visible light photo-catalysis property into ECC for pavement applications. KRONOClean K7000 has a density of 2.9 g/cm^3 with no pigmentary property [32]. The PE fiber has a diameter of 23 μm, a length of 19 mm, a tensile strength of more than 3000 MPa, and an elastic modulus beyond 100 GPa (Table 15.1). The high tensile strength and elastic modulus are achieved due to the use of ultra-high molecular weight PE resin through a gel spinning process during the manufacturing of the PE fiber.

White silica sand is engaged to enhance the skid resistance while maintaining the surface brightness. The white silica sand has a density of 2.65 g/cm^3, a hardness of 7 Mohs, and a maximum particle size of 2.38 mm. Figure 15.1 shows the gradation curve of the white silica sand used in the current study.

15.2.3 Mix Design and Specimen Preparation

Different amounts of silica sand (10, 20, 30%) and T_iO_2 (1, 2, 5%) were incorporated into the white ECC mix to enhance the skid resistance and to impart the photocatalysis property, respectively (Table 15.2). A reference ECC mixture (S00-T0) without the addition of silica sand and T_iO_2 was also prepared.

ECC mixtures were prepared in a planetary mixer. White cement, silica sand, and T_iO_2 were dried mixed first. Water was then added, followed by the addition of HRWR and VMA and mixed for 3–5 min until the mixture achieved the required consistency. Lastly, the PE fibers were slowly added and mixed for another 3–5 min until the fibers were uniformly dispersed. The fresh mixture was cast into cubes

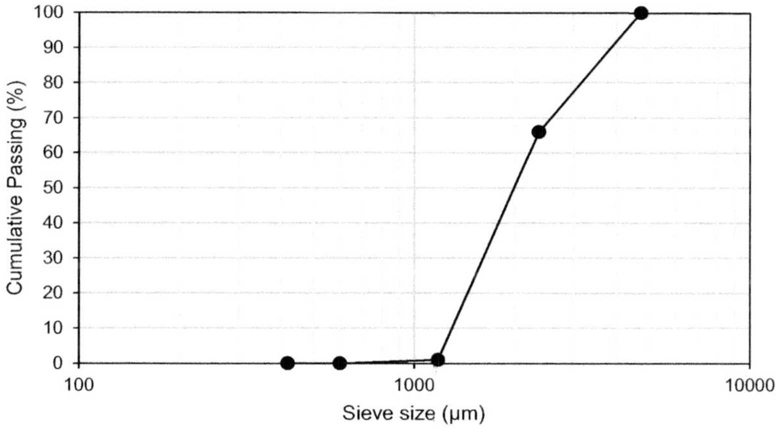

Fig. 15.1 Gradation curve of silica sand

Table 15.2 Mixture design of white ECC with visible light photocatalysis property for pavement applications (kg/m^3)

Mixture	White cement	Silica sand	T_iO_2	Water	HRWRA	VMA (vol.%)	PE fiber (vol.%)
S00-T0	1.0	None	None	0.35	0.02	1.5	1.8
S10-T0	1.0	0.1	None	0.35	0.02	1.5	1.8
S20-T0	1.0	0.2	None	0.35	0.02	1.5	1.8
S30-T0	1.0	0.3	None	0.35	0.03	1.5	1.8
S20-T1	1.0	0.2	0.01	0.35	0.02	1.5	1.8
S20-T2	1.0	0.2	0.02	0.35	0.03	1.5	1.8
S20-T5	1.0	0.2	0.05	0.35	0.03	1.5	1.8

(50 mm by 50 mm by 50 mm) for compressive tests, coupon plates (300 mm by 75 mm by 12 mm) for flexural tests, dog-bone shaped molds with gauge length zone 120 mm (length) by 85 mm (width) tapering to 35 mm with a thickness of 15 mm for uniaxial tensile tests, large plates (300 mm by 300 mm by 10 mm) for skid resistance related tests, and small plates (50 mm by 50 mm by 10 mm) for self-cleaning related tests. Specimens were removed from the mold after one day and were further cured at normal laboratory air (25 ± 3°C, 70%RH) for another 27 days before further testing. For the skid resistance related tests, the top surface of the large plate specimens were treated with a steel brush after the specimens were removed from the mold (i.e., at the age of one day) to expose the silica sand on the surface.

15.2.4 Test Program

The mechanical properties of white ECC incorporating a different amount of silica sand and T_iO_2 were evaluated. The compressive test was carried out following ASTM C109/109 M [3] to determine the compressive strength of the cube specimens at the age of 28 d. The test was done under load control with a loading rate of 25 kN/min. At least four specimens were tested for each mix. The uniaxial tensile test was carried out on dog-bone specimens to determine the tensile behavior of specimens. The test was done under displacement control at 0.0025 mm/s. Two external linear variable displacement transducers (LVTDs) were attached to the dog-bone specimen with a gauge length of 100 mm to measure the specimen deformation. At least three specimens were tested for each mix.

Microscopic skid resistance of white ECC incorporating a different amount of silica sand was evaluated using the British pendulum test (BPT) followed ASTM standard [4]. BPT is a dynamic pendulum impact-type test used to measure the energy loss when the rubber slider slides over the pavement surface. The test results are reported as the British pendulum number (BPN). This method provides a measurement of microtexture and can be used either in the field or laboratory. At least three specimens were evaluated for each mix, and a minimum of five tests were performed for each specimen. The tests were conducted in wet test conditions by spraying water on the specimens' surface. The macroscopic skid resistance was evaluated using the sand patch test (SPT), i.e., a volumetric-based spot test method with a sand patch. The method, which is based on ASTM standard [5], can be used to determine the (average) mean texture depth (MTD) of pavement surface which reflects the macrotexture characteristic of the surface. This method is widely used as it is practical and considerably inexpensive.

Photocatalytic activity of white ECC samples incorporating a different amount of TiO_2 was evaluated by measuring the degradation of dyes under visible light irradiation. Rhodamine B (RhB) dye which shows a strong red colour was used for this test. RhB has been extensively used to study the photocatalytic performance of T_iO_2-based cementitious composites for the following reasons. Firstly, the structure of RhB is similar to the structure of anthracene, an example of polycyclic aromatic hydrocarbons compounds (PAH compounds) [24]. PAHs are an example of soiling agents found in urban environments and mainly produced by incomplete combustion of organic matter. Secondly, RhB is relatively stable in the alkaline environment of the cementitious surface [42]. Lastly, RhB is water-soluble and hence can be spread easily on the surface of cement samples. As the photodecomposition of RhB occurs, the red color will be diminished, and in this case, used as an indication of self-cleaning performance [1, 53]. The procedure is commonly used and standardized in the Italian standard (UNI 11,259:2008). RhB solution was prepared with a concentration of 0.05 g/L. The RhB solution was then applied to the surface of white ECC specimens with a total area of around four cm^2 in a circular shape. The dyed specimens were placed into a chamber equipped with a 0.55 W/m2 visible light lamp for irradiation.

The reflectance of the specimen before and after the different duration of irradiation (i.e., 2 and 4 h) was captured with a UV–Vis spectrophotometer (UV-2600).

Accelerated dirt pick-up resistance (DPUR) of white ECC incorporating a different amount of T_iO_2 was also carried out. The surface brightness of the specimen was measured before and after the application of synthesized dirt, which is a carbon black-based formula synthesized following ASTM D7898-18. The specimens were first sprayed 50 times with the synthesized dirt. The specimens were then rinsed with flowing water for 30 s. The sample surface was dried out, and the brightness was determined with a reflected color measurement spectrophotometer. At least two specimens were tested, and the average brightness is reported for each mixture designs.

15.3 Results

15.3.1 T_iO_2 Characterization

Four commercial T_iO_2 samples were evaluated in this screening test. Figure 15.2 shows the morphologies of the four T_iO_2 nanoparticles, where [9] is the reference to the original paper where the figure appeared. As can be seen, the T_iO_2 particles show a spherical shape with strong flocculation to form larger-size clusters. The primary particles size of P25 and K7000 is estimated to be about 20–40 nm in diameter, while the particle size of T_iO_2-anatase and T_iO_2-rutile is in the range of hundreds nanometer. However, the primary particles of K7000 tend to agglomerate to form larger secondary particles (>500 nm), giving rise to the porous structure.

The specific surface area and pore structure were evaluated by the N2 adsorption experiment. Figure 15.3 presents the physi-sorption isotherm of the four T_iO_2. As can be seen, the physi-sorption isotherm for K7000 shows a hysteresis effect indicating it is mesoporous. In comparison, the observed isotherms for P25, T_iO_2-anatase, and T_iO_2-rutile suggest that these particles are non-porous. The Brunauer–Emmett–Teller (BET) theory was adopted to calculate the surface area (Table 15.3). As can be seen, the calculated BET shows that K7000 has the highest specific surface area of 252 m^2/g followed by P25 (46.9 m^2/g). The high surface area for K7000 was strongly correlated to the meso-pore volume [16]. In contrast, the surface area for T_iO_2-anatase and T_iO_2-rutile samples is only 8 and 10 m2/g, respectively.

The crystal structures of T_iO_2 were investigated through the XRD analysis. Figure 15.4 shows the XRD patterns of P25, K7000, T_iO_2-anatase, and T_iO_2-rutile. As can be seen, XRD analysis confirms that K7000 and T_iO_2-anatase samples consist of anatase phase, while T_iO_2-rutile sample consists of the rutile phase. P25 is the only sample consisting of both anatase and rutile phases, in the ratio of 80–20% (Table 15.3).

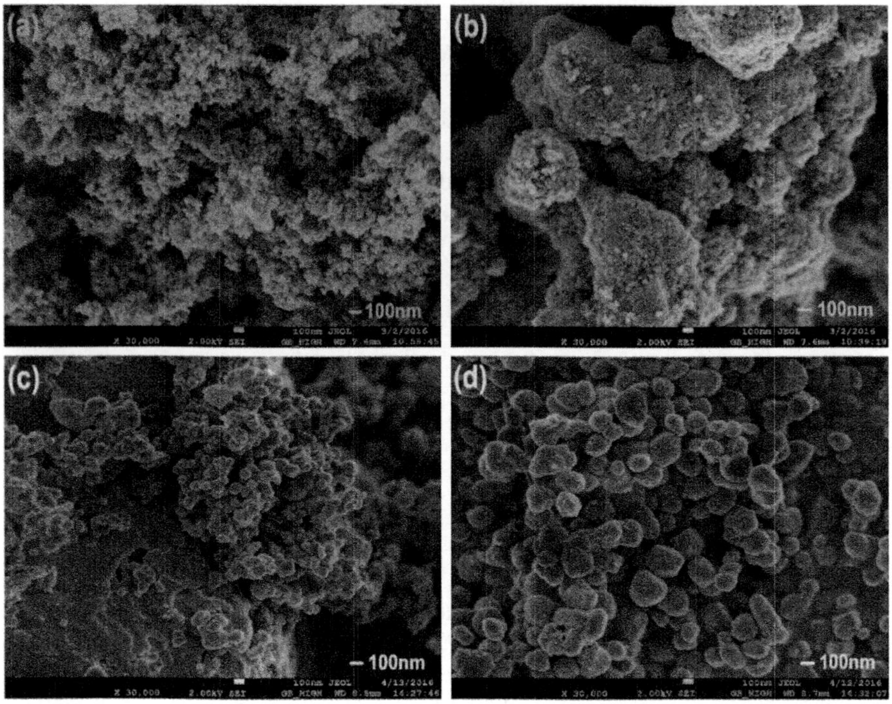

Fig. 15.2 Scanning electron microscope (SEM) images of **a** P25, **b** K7000, **c** T_iO_2-anatase and **d** T_iO_2-rutile

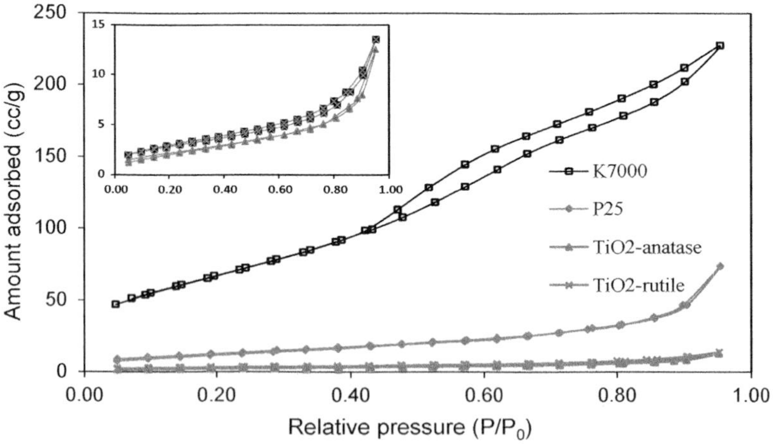

Fig. 15.3 N_2 adsorption isotherms of P25, K7000, T_iO_2-anatase, and T_iO_2-rutile

Table 15.3 Properties of P25, K7000, T_iO_2-anatase, and T_iO_2-rutile

T_iO_2 sample	Crystalline phase	Bandgap energy (eV)	BET-surface area (m^2/g)	Surface charge[a] (mV)	Methylene blue	
					Adsorption (%)	Photo-decomposition (%)
P25	80% Anatase 20% Rutile	3.21	46.9	33.20 ± 0.26	5	21
K7000	Anatase	3.25	252	20.00 ± 0.10	8	43
TiO2-anatase	Anatase	3.20	8	-28.70 ± 2.21	51	17
TiO2-rutile	Rutile	2.95	10	21.13 ± 0.38	6	18

[a]Measured in deionised water. Statistical significance differences (t test): $P < 0.01$

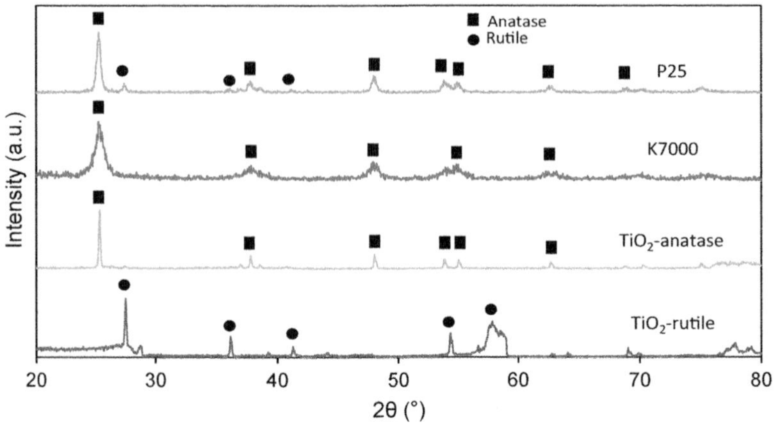

Fig. 15.4 XRD spectra of P25, K7000, T_iO_2-anatase, and T_iO_2-rutile

The optical properties of the four T_iO_2 were examined using diffuse reflectance spectroscopy. Figure 15.5 presents the absorption spectra and bandgap energy calculation of P25, K7000, T_iO_2-anatase, and T_iO_2-rutile. The bandgap of T_iO_2 can be estimated from the absorption spectra using the Kubelka–Munk model and plotted using Tauc's relation.

$$ahv = \text{const}(hv - Eg)^r \tag{15.1}$$

where hv represents the energy of the incident photon and the exponent r denotes the nature of the transition. The Tauc's plot ((αhv)1/r vs. hv) has a distinct linear regime which indicates the onset of absorption. Linear extrapolation of this regime to the energy axis at (αhv)1/r = 0 yields the bandgap energy of a material. Assuming the value of r is 2 (indirect allowed transition) for T_iO_2 [28, 57], the Tauc's plot ((αhv)1/2 vs. hv) is illustrated in Fig. 15.5b. The calculated bandgap of P25, K7000, and T_iO_2-anatase is around 3.2 eV, which that of rutile is slightly less than 3.0 eV (Table 15.3).

Figure 15.6 shows the change in concentration of methylene blue in the aqueous solution as a function of time due to absorption in the first 30 min and photo-catalytic decomposition after the visible light source is on. As can be seen, the dye solution achieved adsorption equilibrium after 15 min under the dark condition. It was found the adsorption capability of P25, K7000, and T_iO_2-rutile is much lower than that of T_iO_2-anatase. The concentration of dye in the solution dropped rapidly in the first 15 min and about 50% of dye was adsorbed on the surface of T_iO_2-anatase.

Figure 15.6 shows that the photocatalytic activity in the presence of T_iO_2-rutile was the lowest one, although T_iO_2-rutile has the smallest bandgap and higher SSA

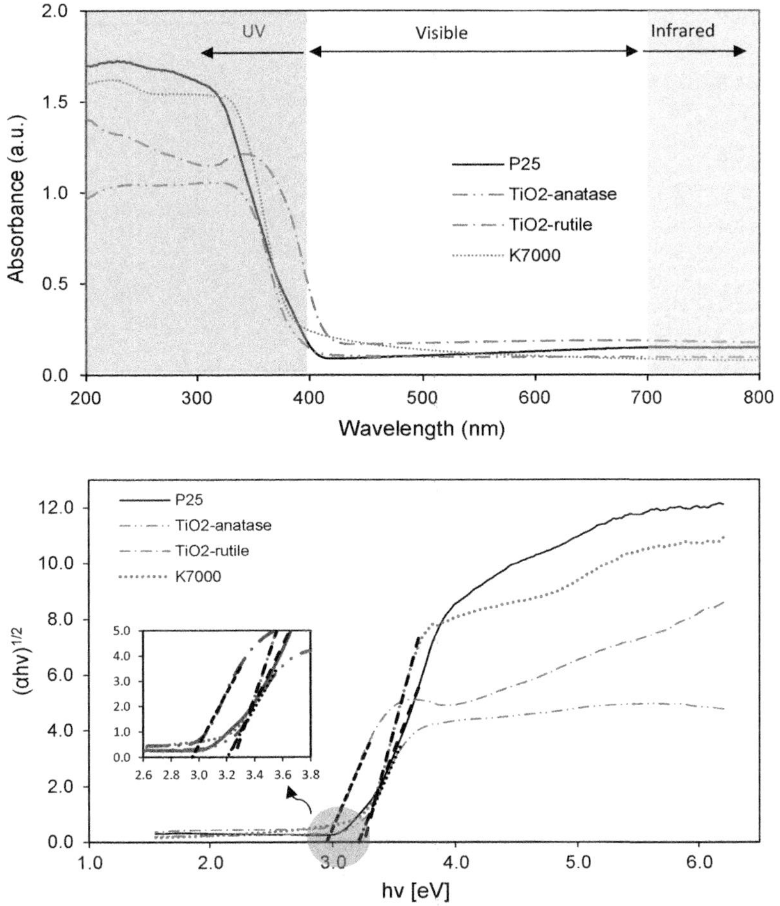

Fig. 15.5 **a** Absorption spectra of P25, K7000, T_iO_2-anatase, and T_iO_2-rutile and **b** the corresponding Tauc's plots

than TiO_2-anatase shown in Table 15.3. It is not surprising as the reaction was thermodynamically not permissible due to the band energy level of T_iO_2-rutile. According to literature, the conduction band of rutile is more positive than the reduction potential of electron acceptor (H + /H2) [14]. In consequence, such a condition inhibits the electron transfer process, which in turn increase electron hole recombination and diminish photocatalytic activity.

It is noteworthy that while T_iO_2-anatase has the lowest surface area according to the BET result (Table 15.3), it shows the highest dye adsorption capability (Fig. 15.6). Previous studies demonstrated that surface charges of T_iO_2 might play an important role in the adsorption process [12, 18, 30]. The electrostatic interactions between the organic dye and T_iO_2 are the important factor controlling the interaction mechanism of contaminant/dye particles to T_iO_2 surfaces [18]. The surface charges of the four

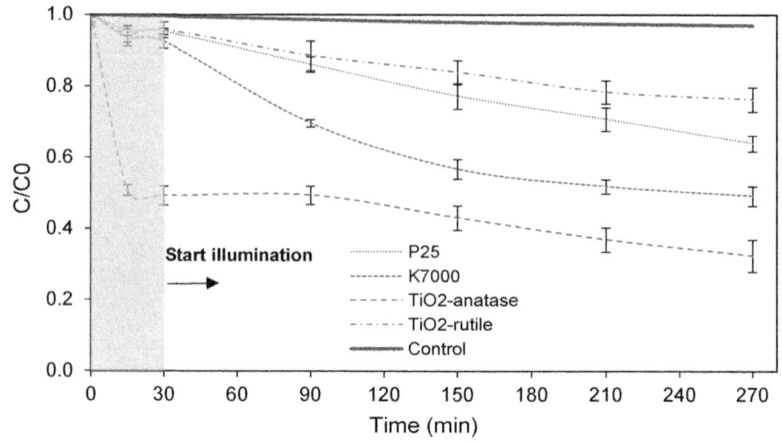

Fig. 15.6 Photo-catalytic performance of the four T_iO_2 in the degradation of methylene blue under visible light irradiation (in solution phase)

T_iO_2 was therefore measured using a zeta potential analyzer. As shown in Table 15.3, the TiO2-anatase was more electro-negative than the other three T_iO_2, supporting stronger adsorption capability with cationic methylene blue dye, which is positively charged.

From Fig. 15.6, it was observed that T_iO_2-anatase displayed the highest total decrease in the concentration of dye after four hours of visible light illumination. However, the decrease of dye concentration for T_iO_2-anatase mainly occurred during the adsorption process (51%, Table 15.3) and was very slow during visible light exposure (17%, Table 15.3). On the contrary, the total decrease in dye concentration for the other commercial products was highly controlled under visible light irradiation. Table 15.3 shows that K7000 (43%) generally outperformed other TiO_2 materials used in this study for the degradation of dye in visible light.

Considering the bandgap of various commercial T_iO_2, it all lies in the UV region, which should have no response in visible light. It was suggested that the degradation of MB dye in visible light was ascribed to the dye photosensitization mechanism [13, 36, 39]. In this procedure, the MB molecule could absorb the visible light to generate electron–hole pairs. The excited electrons are then transferred from a singlet/triplet state of MB (MB*) into the conduction band of T_iO_2, which further could react with oxygen, O_2, to produce O_2- radicals (superoxide anion). The radicals generated could oxidize the aromatic rings of MB. The performance differences among commercial T_iO_2 was mainly caused by their particle size and pore structure, which define the active site available for the photosensitization process. Due to high surface area and porosity, K7000 provide a more active site for reaction, resulting in a good photocatalytic performance.

Due to its high photocatalytic performance under visible light irradiation, K7000 was selected to be used in composite formulation to impart visible light photocatalytic property into white ECC. Further characterization on K7000 was therefore conducted. The chemical state of elements in K7000 was analyzed using XPS analysis. Figure 15.7 shows the full XPS spectrum of T_iO_2 K7000 (Fig. 15.7a) and the corresponding high-resolution XPS spectra and their deconvolution near O 1s (Fig. 15.7b), Ti 2p (Fig. 15.7c), and C 1s (Fig. 15.7d) regions. As can be seen, the main component in the O 1s spectrum is at 531.98 and 533.40 eV, which are attributed to lattice oxygen and surface adsorbed OH group [10]. Next, Ti 2p dominates the spectrum with binding energies of 460.76 and 466.46 eV (Fig. 15.7c). The binding energy at 460.76 eV (Ti 2p3/2) and the binding energy at 466.46 eV (Ti 2p1/2) arises from spin orbit-splitting, which are consistent with Ti4+ in T_iO_2 lattice [49]. Lastly, the deconvoluted C 1s spectrum shows three main peaks at the binding energies of 284.80, 286.70 and 288.23 eV (Fig. 15.7d). Generally, the binding energy of 284.80 eV could be either connected to a signal of adventitious elemental carbon

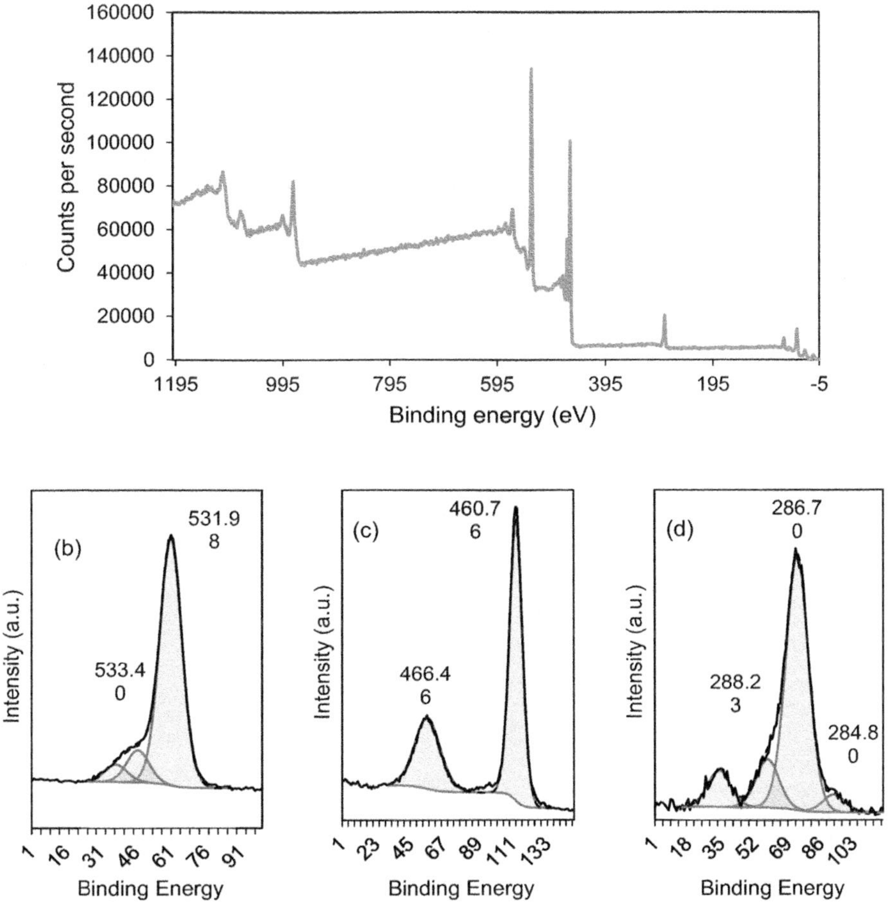

Fig. 15.7 **a** The full XPS spectrum of K7000, and **b–d** high resolution XPS spectra and their deconvolution near **b** O 1 s, **c** Ti 2p, and **d** C 1 s

contamination adsorbed from ambient or ascribed to graphitic C–C bonds [17, 36]. The binding energy of 286.70 and 288.23 eV demonstrated the presence of C–O and C=O bonds as in other works [56, 57]. This tells that the component at those value is usually carbonate-like species in carbon-doped T_iO_2 samples [49]. Besides its high specific surface area, which provides a high number of the active site on the photocatalytic surface, previous study reported that carbon modified T_iO_2 like K7000 shows improved charge separation and electron transfer process leading to improved performance under visible light/low light condition.

15.3.2 *Effects of Silica Sand on Mechanical Properties*

Table 15.4 summarizes the mechanical properties of the four white ECC mixes with different silica sand contents. As can be seen, the compressive strength increases while the tensile strength and tensile ductility of white ECC decrease with increasing silica sand dosage. This may be attributed to the inclusion of hard silica sand increases the matrix toughness, and thus the chances for the formation of steady-state cracks and multiple cracking are reduced [34, 43, 50]. However, all mixes still possess moderate mechanical properties (i.e., compressive strength >30 MPa, tensile strength >3 MPa, and tensile strain capacity >1%) suitable for pavement applications [8, 40].

BPN results of white ECC specimens are presented in Table 15.5. As can be seen, without the addition of silica sand, S00-T0 has a BPN of 54 ± 6.6 lower than the required skid resistance of 55 for new roadways according to the local standard set by the Land Transport Authority of Singapore (LTA [38]. This suggests conventional

Table 15.4 Mechanical properties of ECC

Specimen	Compressive strength (MPa)	Tensile strength (MPa)	Tensile strain capacity (%)
S00-T0	38.06 ± 0.06	5.04 ± 0.75	3.71 ± 0.50
S10-T0	38.56 ± 1.00	3.84 ± 0.51	2.80 ± 0.52
S20-T0	49.02 ± 1.74	3.14 ± 0.79	2.35 ± 0.10
S30-T0	49.10 ± 2.86	3.03 ± 0.16	1.40 ± 0.34
S20-T1	55.07 ± 3.02	3.36 ± 0.96	2.69 ± 0.34
S20-T2	52.81 ± 4.56	4.58 ± 0.46	3.90 ± 1.83
S20-T5	52.41 ± 1.69	4.69 ± 0.55	4.14 ± 1.20

Table 15.5 Skid resistance performance of ECC-SS-T_iO_2

Specimen	Microtexture (BPN)	Macrotexture, MTD (mm)
S00-T0	54 ± 6.6	0.09 ± 0.002
S10-T0	66 ± 0.8	0.22 ± 0.04
S20-T0	79 ± 1.6	0.37 ± 0.04
S30-T0	83 ± 1.9	0.04

ECC cannot be used for road application [7]. A significant improvement in skid resistance was observed when silica sand is included in the ECC mix (BPN above 65). The S10-T0 possesses a BPN of 66, and S20-T0 possesses a BPN of 79, about 20 and 45% improvement, respectively, compared to the control. However, more silica sand inclusion (i.e., S30-T0) only lead to a very mild improvement of the skid resistance of white ECC.

As shown in Table 15.5, the MTD of S00-T0 is very low, with a surface texture of only 0.09 ± 0.002 mm. An addition of 10% silica sand (S10-T0) remarkably increases the MTD to 0.22 ± 0.04 mm. While the addition of 20% silica sand (S20-T0) further enhance the MTD to 0.37 ± 0.04 mm, continuous increment of silica sand dosage only leads to moderate enhancement of MTD (S30-T0).

15.3.3 Effects of TiO_2 on Mechanical Properties

As shown in Table 66, the addition of T_iO_2 leads to a slight improvement of the compressive strength of white ECC from 49 to 52–55 MPa. The maximum improvement was observed with 1% T_iO_2 addition, which is in line with previous studies [33, 45, 46], due to decreased crystalline $Ca(OH)_2$ content required for C-S–H gel formation and unsuitably dispersed nanoparticles in the cement matrix [41, 45, 54]. Furthermore, tensile strength and tensile ductility of white ECC increase with increasing T_iO_2 content. The addition of T_iO_2 nano-particles may densify the interface between fiber and matrix, leading to an enhanced interface friction bond. As a result, the fiber bridging strength as well as complimentary energy increase, and thus the tensile strength and tensile ductility of white ECC increase.

Photocatalytic activity of white ECC samples incorporating a different amount of T_iO_2 was evaluated by measuring the degradation of RhB after visible light irradiation, as shown in Fig. 15.8. After light exposure, the reflectance spectra exhibit a recovery after two hours of irradiation, and yet even a stronger recovery after four hours irradiation of visible light for samples with T_iO_2 addition. Distinct reflectance spectra were observed in the UV region (<400 nm) after the addition of T_iO_2. The reflectance in the UV region reduces with increasing T_iO_2 dosage. This is because K7000 has strong absorption of UV, as shown in Fig. 15.5. Furthermore, the reflectance spectra in the visible light range (400–700 nm) is slightly reduced for samples (w/o RhB) with T_iO_2 addition due to minor absorption of visible light of K7000 (Fig. 15.5). When RhB was applied to the sample surface (w/ RhB), significant reflectance reduction was observed in the wavelength between 450 and 600 nm with the lowest peak at 555 nm, which is corresponding to the absorption spectrum of the pink color of RhB solution. The application of RhB also affects the spectra in the UV region for the specimen with 0% T_iO_2 (i.e., ST20-T0), as shown in Fig. 15.8a, due to the domination of RhB reflectance at the UV region.

Figure 15.9 shows the reflectance recovery of white ECC with different T_iO_2 content after visible light irradiation. The reflectance recovery was defined as the ratio of the total reflectance recovery after visible light irradiation exposure (i.e.,

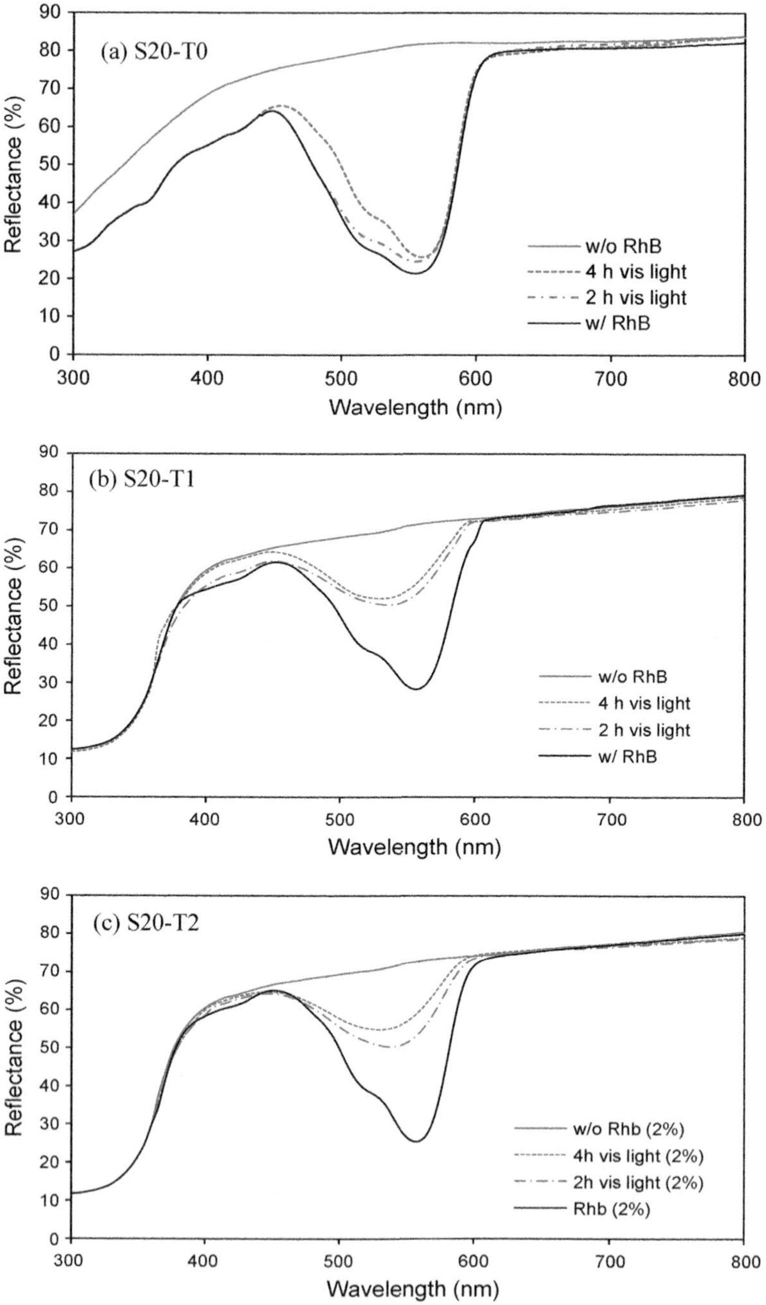

Fig. 15.8 Photo-decomposition performance of **a** S20-T0, **b** S20-T1, **c** S20-T2, and **d** S20-T5 under visible light irradiation

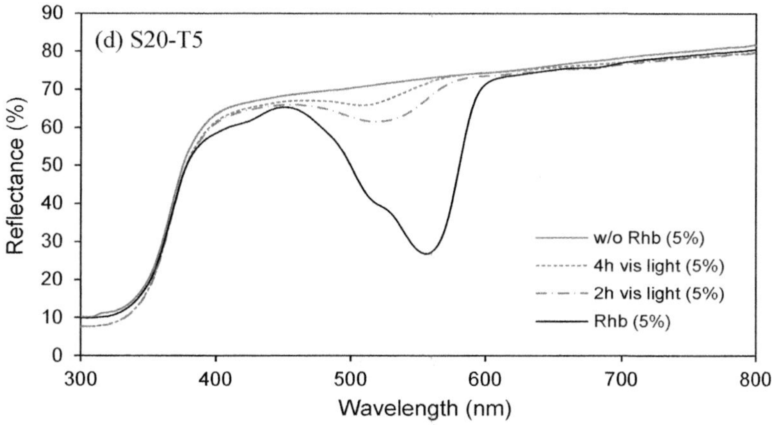

Fig. 15.8 (continued)

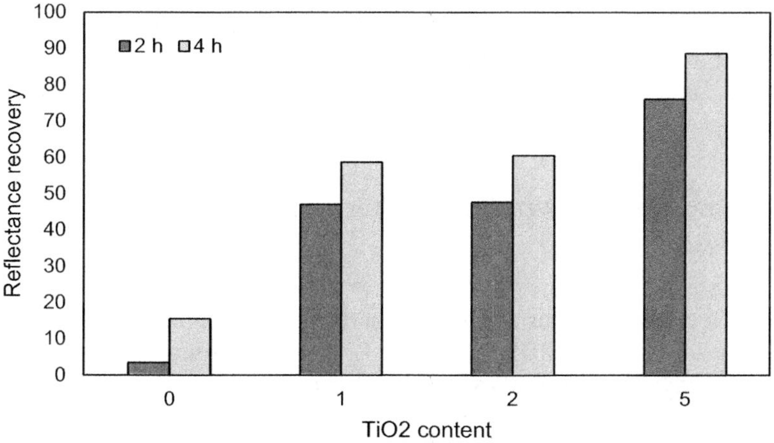

Fig. 15.9 Reflectance recovery of ECC-SS-T_iO_2 based on visible light irradiation

area between the x h visible light curve and the w/RhB curve) to the total reflectance reduction after applying RhB (i.e., area between the w/o RhB curve and the w/RhB curve).

$$Solar\ reflectance\ recovery\ (\%) = \frac{Area\ between\ curve(\times\ h\ vis)\ and\ curve\left(\frac{w}{RhB}\right)}{Area\ between\ curve\ (\times h\ vis)\ and\ curve\left(\frac{w}{RhB}\right)} \times 100 \quad (15.2)$$

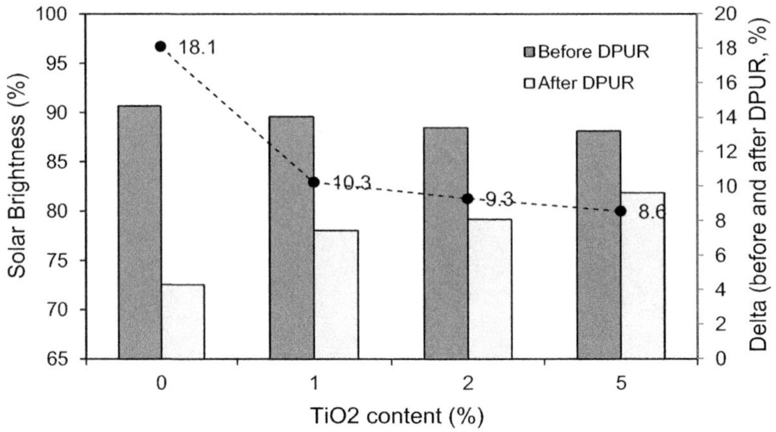

Fig. 15.10 Accelerated DPUR under visible light

As can be seen, reflectance recovery increases with increasing T_iO_2 dosage and visible light exposure duration. With 5% T_iO_2 addition, the reflectance recovery of white ECC can reach up to almost 89% after four hours of visible light irradiation.

15.3.4 Accelerated Dirt-Pick up Resistance (DPUR) Under Visible Light

Figure 15.10 presents the surface brightness (L*) of the white ECC samples before and after the accelerated DPUR test, where L* = 100 represents the brightest white and L* = 0 indicates the darkest black. As can be seen, the surface brightness decreases with increasing T_iO_2 dosage due to the minor absorption of the visible light of K7000. The surface brightness of all white ECC samples were reduced after the accelerated DPUR test. However, samples with higher T_iO_2 content showed less reduction of the surface brightness. With 5% T_iO_2 addition, the surface brightness reduction of S20-T5 is only about 8%. This means dirt is more difficult to stick to the surface on the resulting white ECC.

References

1. Ahmad I, Kan C (2017) Visible-light-driven, dye-sensitized TiO_2 photo-catalyst for self-cleaning cotton fabrics. Coatings 7(11):192. https://doi.org/10.3390/coatings7110192
2. Andaloro A, Mazzucchelli ES, Lucchini A, Pedeferri MP (2017) Photocatalytic self-cleaning coatings for building facade maintenance. Performance analysis through a case-study application. FDE 4(3–4):115–129. https://doi.org/10.3233/FDE-160054

3. ASTM C109 (2020) Standard test method for compressive strength of hydraulic cement mortars. ASTM International, West Conshohocken, PA. https://doi.org/10.1520/C0078_C0078M-18
4. ASTM E303–93 (1993) ASTM standard test method for measuring surface frictional properties using the British pendulum tester. ASTM International, West Conshohocken, PA.
5. ASTM E965 - 96 (2006) Standards test method for Measuring Pavement Macrotexture Depth Using a volumetric technique (reapproved 2006). ASTM International, West Conshohocken, PA.
6. Awadalla A, Zain MFM, Kadhum AA, Abdalla Z (2011) Improve indoor air quality by using titanium dioxide as coating photocatalyses under UV irradiation. Res J Appl Sci 6(2):99–103. https://doi.org/10.3923/rjasci.2011.99.103
7. Bawono AA, NguyenDinh N, Yang EH, Lechner B (2017) Evaluation of highway pavement skid resistance performance made of Engineered Cementitious Composite (ECC). In: Tenth international conference on the bearing capacity of roads, railways and airfields, Athens. CRC Press, Greece
8. Bawono AA, Lechner B, Yang E-H (2019) Skid resistance and surface water drainage performance of engineered cementitious composites for pavement applications. Cement Concr Compos 104. https://doi.org/10.1016/j.cemconcomp.2019.103387
9. Bawono AA, Tan ZH, Hamdany AH, NguyenDinh N, Qian S, Lechner B, Yang E-H (2020) Bright and slip-proof engineered cementitious composites with visible light activated photocatalysis property for pavement in tunnels. Cement Concr Compos 114. https://doi.org/10.1016/j.cemconcomp.2020.103788
10. Bharti B (2016) Formation of oxygen vacancies and Ti3+ state in TiO_2 thin film and enhanced optical properties by air plasma treatment. Sci Rep 6
11. Boonen E, Akylas V, Barmpas F, Boréave A, Bottalico L, Cazaunau M, Chen H, Daële V, de Marco T, Doussin JF, Gaimoz C, Gallus M, George C, Grand N, Grosselin B, Guerrini GL, Herrmann H, Ifang S, Kleffmann J, Kurtenbach R, Maille M, Manganelli G, Mellouki A, Miet K, Mothes F, Moussiopoulos N, Poulain L, Rabe R, Zapf P, Beeldens A (2015) Construction of a photocatalytic de-polluting field site in the Leopold II tunnel in Brussels. J Environ Manage 155:136–144. https://doi.org/10.1016/j.jenvman.2015.03.001
12. Butman MF, Ovchinnikov NL, Karasev NS, Kochkina NE, Agafonov AV, Vinogradov AV (2018) Photocatalytic and adsorption properties of TiO2-pillared montmorillonite obtained by hydrothermally activated intercalation of titanium polyhydroxo complexes. Beilstein J Nanotechnol 9:364–378. https://doi.org/10.3762/bjnano.9.36
13. Cabir B, Yurderi M, Caner N, Agirtas MS, Zahmakiran M, Kaya M (2017) Methylene blue photocatalytic degradation under visible light irradiation on copper phthalocyanine-sensitized TiO_2 nanopowders. Mater Sci Eng, B 224:9–17. https://doi.org/10.1016/j.mseb.2017.06.017
14. Carp O, Huisman C, Reller A (2004) Photoinduced reactivity of titanium dioxide. Prog Solid State Chem 32(1–2):33–177. https://doi.org/10.1016/j.progsolidstchem.2004.08.001
15. Cassar, L., Pepe, C., Pimpinelli, N., Tognon, G., Guerrini, G., Amadelli, R. (2003). White cement for architecture concrete, possessing the photocatalytic properties.
16. Chang S-S, Clair B, Ruelle J, Beauchêne J, Di Renzo F, Quignard F, Zhao G-J, Yamamoto H, Gril J (2009) Mesoporosity as a new parameter for understanding tension stress generation in trees. J Exp Bot 60(11):3023–3030. https://doi.org/10.1093/jxb/erp133
17. Chen C, Cai W, Long M, Zhou B, Wu Y, Wu D, Feng Y (2010) Synthesis of visible-light responsive graphene oxide/TiO(2) composites with p/n heterojunction. ACS Nano 4(11):6425–6432. https://doi.org/10.1021/nn102130m
18. Dai M (1994) The effect of zeta potential of activated cabon on the adsorption of dyes from aquesous solutin. J Colloid Interface Sci 164:223–228
19. European Asphalt Pavement Association (2008) Asphalt pavement in tunnels, Belgium
20. Fotios S, Boyce P, Ellis C (2005) The effect of pavement material on road lighting performance. PPAD 9/100/77, Sheffield
21. Fujishima A, Honda K (1972) Electrochemical photolysis of water at a semiconductor electrode. Nature 238:37–38

22. Fujishima A, Zhang X, Tryk D (2008) TiO_2 photocatalysis and related surface phenomena. Surf Sci Rep 63(12):515–582. https://doi.org/10.1016/j.surfrep.2008.10.001
23. Gallus M, Akylas V, Barmpas F, Beeldens A, Boonen E, Boréave A, Cazaunau M, Chen H, Daële V, Doussin JF, Dupart Y, Gaimoz C, George C, Grosselin B, Herrmann H, Ifang S, Kurtenbach R, Maille M, Mellouki A, Miet K, Mothes F, Moussiopoulos N, Poulain L, Rabe R, Zapf P, Kleffmann J (2015) Photocatalytic de-pollution in the Leopold II tunnel in Brussels: NOx abatement results. Build Environ 84:125–133. https://doi.org/10.1016/j.buildenv.2014.10.032
24. Gustafson KE, Dickhut RM (1996) Particle/gas concentrations and distributions of PAHs in the atmosphere of Southern Chesapeake Bay. Environ Sci Technol 31(1):140–147
25. Hall J, Smith K, Titus-Glover L, Wambold J, Yager T, Rado Z (2009) Guide for pavement friction, Washington, D.C
26. Henry J (2000) Evaluation of pavement friction characteristics: a synthesis of highway practice, Washington, D.C
27. Jackett MJ, Frith WJ (2009) Measurement of the reflection properties of road surfaces to improve the safety and sustainability of road lighting. NZ Transport Agency research report, vol 383. NZ Transport Agency, Wellington, New Zealand
28. Jimenez-Relinque E, Llorente I, Castellote M (2017) TiO_2 cement-based materials: understanding optical properties and electronic band structure of complex matrices. Catal Today 287:203–209. https://doi.org/10.1016/j.cattod.2016.11.015
29. Jofré C, Romero J, Rueda R (2010) Contribution of concrete pavements to the safety of tunnels in case of fire, Brussels
30. Karaca S, Gurses A, Açıkyıldız M (2008) Adsorption of cationic dye from aqueous solutions by activated carbon. Microporous Mesoporous Mater 115(3):376–382
31. Knaze P (20015) Concrete pavements in tunnels, Bratislava
32. KRONOS (2011) KRONOClean7000 degrades pollutants with visible light and with UV radiation. https://kronostio2.com/en/
33. Kushwaha AS, Saxena R, Pal S (2015) Effect of titanium dioxide on the compressive strength. J Civ Eng Environ Technol 2(6):482–486
34. Li J, Yang E-H (2017) Macroscopic and microstructural properties of engineered cementitious composites incorporating recycled concrete fines. Cement Concr Compos 78:33–42. https://doi.org/10.1016/j.cemconcomp.2016.12.013
35. Linsebigler AL, Lu G, Yates JT (1995) Photocatalysis on TiO2 surfaces: principles, mechanisms, and selected results. Chem Rev 3(3):735–758
36. Liu J, Zhu W, Yu S, Yan X (2014) Three dimensional carbogenic dots/TiO2 nanoheterojunctions with enhanced visible light-driven photocatalytic activity. Carbon 79:369–379. https://doi.org/10.1016/j.carbon.2014.07.079
37. Longley I (2014) Criteria for in-tunnel and ambient air quality, New South Wales
38. LTA Singapore (2018) Code of practice traffic control at work zone. LTA Singapore, Singapore
39. Mamba G, Mbianda XY, Mishra AK (2014) Gadolinium nanoparticle-decorated multiwalled carbon nanotube/titania nanocomposites for degradation of methylene blue in water under simulated solar light. Environ Sci Pollut Res Int 21(8):5597–5609. https://doi.org/10.1007/s11356-014-2499-2
40. Mueller S, Mechtcherine V (2017) Use of strain-hardening cement-based composites in real scale applications. In: 4th international RILEM conference on strain hardening cement-based composites, Dresden, Germany, 18–20 September. Springer, Dordrecht
41. Pacheco-Torgal F, Jalali S (2011) Nanotechnology: Advantages and drawbacks in the field of construction and building materials. Constr Build Mater 25(2):582–590. https://doi.org/10.1016/j.conbuildmat.2010.07.009
42. Ruot B, Plassais A, Olive F, Guillot L, Bonafous L (2009) TiO2-containing cement pastes and mortars: measurements of the photocatalytic efficiency using a rhodamine B-based colourimetric test. Sol Energy 83(10):1794–1801. https://doi.org/10.1016/j.solener.2009.05.017
43. Sahmaran M, Lachemi M, Hossain KM, Ranade R, Li V (2009) Influence of aggregate type and size on ductility and influence of aggregate type and size on ductility and mechanical properties of Engineered Cementitious Composites. ACI Mater J 106(3):308–316

44. Salata F, Golasi I, Bovenzi S, Vollaro E, Pagliaro F, Cellucci L, Coppi M, Gugliermetti F, Vollaro A (2015) Energy optimization of road tunnel lighting systems. Sustainability 7(7):9664–9680. https://doi.org/10.3390/su7079664
45. Silvestre J, Silvestre N, de Brito J (2015) Review on concrete nanotechnology. Eur J Environ Civ Eng 20(4):455–485. https://doi.org/10.1080/19648189.2015.1042070
46. Sorathiya J, Shah S, Kacha S (2017) Effect on addition of nano Titanium Dioxide TiO_2 on compressive strength of cementitious concrete. In: International conference on research and innovations in science, engineering & technology, Anand, India
47. Synnefa A, Karlessi T, Gaitani N, Santamouris M, Assimakopoulos DN, Papakatsikas C (2011) Experimental testing of cool colored thin layer asphalt and estimation of its potential to improve the urban microclimate. Build Environ 46(1):38–44. https://doi.org/10.1016/j.buildenv.2010.06.014
48. The World Health Organization (2000) Guidelines for air quality. The WHO, Geneva
49. Tobaldi DM, Seabra MP, Otero-Irurueta G, de Miguel YR, Ball RJ, Singh MK, Pullar RC, Labrincha JA (2015) Quantitative XRD characterisation and gas-phase photocatalytic activity testing for visible-light (indoor applications) of KRONOClean 7000®. RSC Adv 5(124):102911–102918. https://doi.org/10.1039/C5RA22816F
50. Ueda N, Kawamoto A (2017) Influence of coarse aggregate on the mechanical properties. Strain-hardening cement-based composites. In: 4th international RILEM conference, Dresden, Germany, 18–20th September. Springer
51. Vercammen J, Audenaert A (2016) High performance lighting—on the road for energy efficiency in tunnels. In: 8th international conference 'tunnel safety and ventilation, Graz
52. Witkowski H, Jackiewicz-Rek W, Chilmon K, Jarosławski J, Tryfon-Bojarska A, Gąsiński A (2019) Air purification performance of photocatalytic concrete paving blocks after seven years of service. Appl Sci 9(9):1735. https://doi.org/10.3390/app9091735
53. Yang E-H, Garcez E, Li VC (2012) Development of pigmentable engineered cementitious composites for architectural elements through integrated structures and materials design. Mater Struct 45(3):425–432. https://doi.org/10.1617/s11527-011-9774-1
54. Yousefi A, Allahverdi A, Hejazi P (2013) Effective dispersion of nano-TiO_2 powder for enhancement of photocatalytic properties in cement mixes. Constr Build Mater 41:224–230. https://doi.org/10.1016/j.conbuildmat.2012.11.057
55. Zhang H, Keoleian G, Lepech M (2008) An integrated life cycle assessment and life cycle analysis model for pavement overlay systems. In: First international symposium on life-cycle civil engineering, Varenna, Lake Como, Italy, June 10–14. CRC Press, Boca Raton
56. Zhang H, Lv X, Li Y, Wang Y, Li J (2010) P25-graphene composite as a high performance photocatalyst. ACS Nano 4(1):380–386. https://doi.org/10.1021/nn901221k
57. Zhang Y, Pan C (2011) TiO_2 graphene composite from thermal reaction of graphene oxide and its photocatalytic activity in visible light. J Mater Sci 46(8). https://doi.org/10.1007/s10853-010-5116-x
58. Zhao A, Yang J, Yang E-H (2015) Self-cleaning engineered cementitious composites. Cement Concr Compos 64:74–83. https://doi.org/10.1016/j.cemconcomp.2015.09.007

Chapter 16
Conclusions and Recommendations

This chapter summarizes the results and analysis obtained in this thesis. The thesis proposes a novel comprehensive study of the functional performance of engineered cementitious composites (ECC) for pavement application with a case study: electrified roadway with dynamic wireless power transfer in Singapore.

Based on the literature, the conventional concrete pavement design may not be suitable for the electrified roadway with inductive charging. It uses many steel-reinforcements, which potentially increasing losses in power transfer efficiency. Therefore, a new pavement design needs to be developed to overcome this issue. In this thesis, The ECC material has been carefully studied for an innovative precast electrified roadway design. By using a method of a modified Performance-Driven Design Approach (PDDA), the ECC has been successfully modified to improve the functional performance without neglecting the structural performance (Chap. 6). Laboratory and field tests have been conducted to study the skid resistance (Chaps. 7, 8), surface water drainage (Chaps. 10, 11), and noise (Chap. 14). Validated Finite Element and Fluid Dynamic models have also been utilized in this thesis to understand better about the functional performance for roadway made of ECC (Chaps. 9, 12). A comprehensive multi-criteria decision analysis (MCDA) is then developed in this thesis to have a better understanding of the standards, the surface treatment methods, and the measurement method (Chap. 5). More than 10,000 data points have been collected worldwide and analyzed to identify the performance for each surface treatment. The MCDA gives a comprehensive overview on how the surface pavement is designed which then provides better understanding in investigating as well as designing the functional performance, especially in skid resistance, surface water drainage, and noise, of pavement made of ECC. The steps are illustrated in flow charts (see Appendix B).

According to the laboratory results conducted in this study, a conventional ECC does not fulfil the safety requirements for roadways application as it produces a low skid resistance performance. A modified mixture design of ECC for roadway application is then developed in this thesis by introducing corundum and silica sand

A. A. Bawono, *Engineered Cementitious Composites for Electrified Roadway in Megacities*, Springer Theses, https://doi.org/10.1007/978-3-030-88542-7_16

material into ECC, namely ECC-Cor and ECC-SS. The results shown the skid resistance and surface water drainage performance of the modified ECC is significantly improved (Chaps. 1–9). Yet, it was not sufficient as the modified ECC did not provide sufficient surface drainage for the water outflow. Further surface treatment by introducing grooves is successfully improved the water outflow (Chaps. 10–12). A strong correlation model between skid resistance and water outflow according to pavement geometry is proposed in this thesis (Chap. 13).

In this thesis, the tire-pavement noise performance on the pavement made of ECC is predicted with an innovative method. The texture of the slab is measured by using a laser scanner, then analyzed by finite element analysis with SPERoN. Further, the texture of ECC-Cor is modified by introducing the grooves. Results have shown that the sound level has been successfully reduced and that it met the requirements set by the roadway authorities (Chap. 14).

Additional innovative functions, i.e., self-cleaning and air purification for ECC, embedded in the pavement is also proposed in this thesis (Chap. 15). Self-cleaning helps to maintain the brightness and reflectivity of the surface, while air purification improves the air quality in urban areas and in tunnels. The inclusion of TiO_2 in ECC-SS successfully engages photocatalytic function under the visible-light that facilitates the decomposition of Rhodamine B (RhB). The addition of TiO_2 increases self-cleaning performance.

Further, in this final chapter, key contributions, findings, and recommendation for further research according to the specific topic of functional performance on ECC material for pavement application are summarized for each topic.

16.1 The Study of MCDA

Key contributions and findings from the study include:

- The study of MCDA is carefully designed to understand better about the functional performance of pavement. The MCDA method consists of many steps with two main key steps being: (1) scoring, and (2) weighting. The new scoring criteria is set up based on literature review regulated by authorities. The criteria are set from 1 to 5 (1 = lowest performance, and 5 = highest performance), and they are mainly based on the performance, required actions, and risk of accidents. The new weight proportion has been proposed for specific MCDA study on a functional performance for surface pavement with a prioritized safety function. The criteria studied includes durability, cost, microtexture, macrotexture, surface drainage, tire-pavement noise, and pavement roughness.
- The second analysis is done by plotting multiple constraints in the scenario. This analysis allows comparison for many texturing methods. However, this analysis is limited to two constraints. The plot is not determined to give a correlation between constraint (criterion), instead it gives a picture what are the performance according to the two constraints for each texturing method.

- More than about 10,000 data points have been collected, and then they are analyzed for different texturing methods for asphalt and concrete pavements for seven different criteria (constraint): microtexture and friction, macrotexture, water outflow, pavement roughness, and tire-pavement noise, cost, and durability.
- According to the results, the best practice of texturing methods for skid resistance and macrotexture includes longitudinal diamond grooving, transverse diamond grooving, and turf drag. Additionally, exposed aggregate, longitudinal tining, and burlap drag can be used to improve skid resistance to the required safety standards.
- In terms of surface drainage, diamond grooving gives the best results. However, other texturing methods such as diamond grinding, tining, burlap drag, exposed aggregate resulting in sufficient performance for water outflow.
- The optimum texturing methods for both low tire-pavement noise and skid resistance includes longitudinal tining, turf drag, broom drag, exposed aggregate, and diamond grinding. The least effective solution for noise is transverse diamond grooving.

Recommendations for further research on MCDA are given:

- To include more standards/regulations/methods which are existing worldwide.
- To update the questionnaires on the scoring and weighting based on current policies. As policies might be changing from time to time. Alternatively, scoring and weighting can be opted to be changed by the user. In this way, the user can define their own MCDA based on their local standards/regulations.
- To develop more robust software tools/websites so many stakeholders can generate information based on data collected and analyzed.

16.2 The Study of Skid Resistance

Key contributions and findings from the study include:

- A normal ECC (ECC-Cor 0) has a British Pendulum Number (BPN) of 53 ± 9.1 which is lower than the required skid resistance of 55 for new roadways in accordance with local standard set by the Land Transport Authority of Singapore [1]. This suggests ECC without any modification (ECC-Cor 0) cannot be used for road application unless surface treatments are introduced.
- ECC mixed with corundum (i.e., ECC-Cor 50 to 100) with surface exposure after three days shows some improvement in skid resistance as most of the results are above 55 PN on average, except ECC-Cor 90. The moderate improvement may be attributed to delayed surface exposure process at day 3, where the matrix, a fiber-reinforced paste, was very hard to be removed and thus less corundum aggregates were exposed on the surface.
- A significant improvement of skid resistance of ECC-Cor with an exposing process one day after casting was observed (BPN above 65). It can be observed that skid resistance (BPN) increases with increasing amount of corundum used in the ECC-Cor mix. The ECC-Cor 100 possesses a BPN of 75, about 40% improvement compared to ECC-Cor 0. This is because the surface exposure process was

carried out at day 1 and the paste can be removed easily and evenly as evidenced by the reduced variation of the BPN results.

- High skid resistance is also observed for the ECC-Cor specimens with exposed method (BPN around 75, comparable to the ECC-Cor 100). Even with much less corundum was included in the ECC-Cor 0-S specimen (150 kg/m^3) as compared to other ECC-Cor mixes (e.g., 225 and 450 kg/m^3 for ECC-Cor 50 and 100 mixes, respectively). Since skid resistance is a surface property, ECC-Cor 0-S represents a potential approach to strategically use corundum to improve the surface property while keeping the superior mechanical performance of the original ECC. Ultimately, this approach may be applied for both cast-in-situ and precast construction.
- A significant improvement of skid resistance was observed when silica sand is included in the ECC mix (BPN above 65). The ECC-S10 possesses a BPN of 66 and S20 possesses a BPN of 79, about 20 and 45% improvement, respectively, compared to the control. However, more silica sand inclusion (i.e., S30-T0) only lead to very mild improvement of the skid resistance of white ECC.
- The Mean Texture Depth (MTD) of ECC-S00 is very low with the surface texture of only 0.09 ± 0.002 mm. Addition of 10% silica sand (ECC-S10) remarkably increases the MTD to 0.22 ± 0.04 mm. The addition of 20% silica sand (ECC-S20) further enhance the MTD to 0.37 ± 0.04 mm, continuous increment of silica sand dosage only leads to moderate enhancement of MTD (ECC-S30).
- BPN for longitudinal grooves is considered to remain the same, while BPN for transversal grooves increases. MTD for ECC with grooves is also increased about three times compare to specimen without grooves.
- The grooved ECC mixed with aggregates method comes with its advantages including: (1) it significantly increases the skid resistance, and (2) it is relatively easy procedure to cast ECC mixed with aggregates followed by grooving process. The grooves can be done by both precast and groove in situ methods.
- The results from casting in situ ECC in large scale does not satisfy the requirement for pavement application. Not only is the skid resistance considerably low, but it also results in wrinkles. The wrinkles occurred due to the low workability of the ECC itself. The existence of fiber inside the mixture combined with low water cement ration caused difficulties when levelling works needed to be done in the field. Therefore, material ECC is not recommended for pavement application, except there is an improvement on the mixture design to improve the workability for the material.
- More slabs of modified ECC were made with the precast method with steel formworks. By using steel formworks, it eliminates the deformation which was occurred on the first trial with wooden formwork. About 12 slabs were mixed and casted in different days in Tuas South Avenue area. Five full-size slabs and 6 half-size slabs were installed at a busbay in Jurong Island. BPT and SPM tests were taken on site. Various results are observed due to differences during mixing, casting, and finishing work. However, all slabs with precast method result more than 55 BPN, with an average of 68 ± 6.5 BPN for slab 02 to slab 11. The average for macrotexture is 0.44 ± 0.09 mm.

- Alternatively, the reverse precast method is designed to solve previous issues faced including deformation in the early stage, as well as inconsistencies in flat and smooth surfaces. The reverse precast method for modified ECC gives results in microtexture 68 ± 2.45 BPN and macrotexture MTD 0.24 ± 0.02 as average value. Further, this method improves the end-result of physical appearance as the surface is flatter and smoother compared to other previous methods.
- The trend shows BPN reduces while MTD remains unchanged with increasing load cycles. Even after 100,000 load cycles, the BPN is still above 65 BPN and the MTD is maintained at 0.34 mm. It can be concluded the surface texture and skid resistance made of ECC-Cor can last until the end of the pavement life-time without any further intervention. This is mainly due to the hardness of corundum.
- The 3D FEM model on BPT for skid resistance was conducted to predict the BPN of ECC samples. Validation and calibration of the model were performed with consideration of key parameters of the BPT mechanism: the mass, the center of gravity of the model and the sliding distance of the rubber slider. The simulation results show a good correlation with laboratory measurements.
- The 3D FEM model was validated by comparing the BPN results from the FEM analysis with BPN results from laboratory tests. The sample used for validation is from sample with mixture design of ECC-Cor 80. Samples were prepared for both flat and grooved texture. From the results, it can be seen that the difference between FEM model and laboratory test is below 5%, a considerably good model.

Recommendations for further research are given:

- The laboratory scale test on durability shows a good trend on durability. A further study of the durability on skid resistance study based on field investigation can be performed every after 10 years to investigate the long-term performance.
- Further study on skid resistance based on different textures can be performed. The utilization of a silicone-based method for the formwork can be easily applied to study different textures.

16.3 The Study of Surface Water Drainage

Key contributions and findings from the study include:

- The outflow times on normal ECC cannot be measured as the surface is very smooth, resulting in the rubber ring of the Outflow test (OFT) meter device sticking tightly to the surface. This condition makes water flowing out at a very slow rate. The outflow times on ECC mixed with corundum is improved as the corundum provides a microtexture on the surface pavement which allowing the water to find a gap between the aggregates. The average outflow time result of ECC-Cor is 62.11 ± 4.99 s. A similar result is observed in ECC-SS 01 with 65.06 ± 3.51 s.
- The outflow times for the grooved ECC-Cor specimens are even shorter. The combination of the microtexture provided by the corundum and the macrotexture provided by the grooves makes the space between the rubber ring of the Outflow

Meter device and the surface pavement specimens even bigger, which results in a lower value of outflow times. The outflow times of grooved ECC-Cor with a land area of 25.0 ± 0.5, 20.0 ± 0.5, 15.0 ± 0.5, and 7.5 ± 0.5 mm is 7.73, 7.30, 5.83 and 3.38 s, respectively.

- The average Water film thickness (WFT) on ECC-Cor for water drained above the ECC-Cor surface pavement (grade 1%) with flow rates of 1.0, 3.1, 5.1, and 8.9 L/min, are 1.33, 1.91, 2.66, and 4.10 mm, respectively. There is not much difference between the average WFT results derived from the equation and direct measurement (less than 10%). The average WFT on ECC-Cor for water drained above the ECC-Cor surface pavement (grade 5%) with flow rates of 1.0, 3.1, 5.1, and 8.9 L/min, are 1.03, 1.18, 1.93, and 2.78 mm, respectively. There is not much difference between the average WFT results derived from the equation and direct measurement (less than 10%).
- The average WFT on grooved ECC-Cor for water drained above the ECC-Cor surface pavement (grade 1%) with flow rate of 1.0, 3.1, 5.1, and 8.9 L/min, are 0.09, 0.15, 0.22, and 0.35 mm, respectively. There is not much different between the average WFT results derived from the equation and direct measurement (less than 10%). The average WFT on grooved ECC-Cor for water drained above the ECC-Cor surface pavement (grade 5%) with flow rate of 1.0, 3.1, 5.1, and 8.9 L/min, are 1.03, 1.18, 1.93, and 2.78 mm, respectively. There is not much different between the average WFT results derived from the equation and direct measurement (less than 10%).
- The WFT will be significantly decreased when such a surface pavement is equipped with grooves as the water will fill the grooves immediately when it drained above the surface. Once, the grooved filled with water, the remaining water will then only drain along the original surface of the pavement.
- The grooves can reduce the WFT by up to 50% of its original thickness, depending on the dimension of the grooves, as well as the grade of the pavement.
- When the grade of the slab is increased, the average WFT will be reduced. The higher the grade, the greater the reduction of the WFT. With a similar dimension of grooves, the WFT is reduced up to 30% between the flat and grooved ECC-Cor with 1% grade while the WFT can be reduced up to 50% between the flat and grooved ECC-Cor with 5% grade.
- From field investigation, it was found that the outflow results of brushed ECC mixed with corundum improved when there is more corundum on the surface as they have been exposed (see after six hours and 21 h). When the matrix is not fully hardened, more corundum aggregated have been removed away in the process.
- The difference between time of processing does not affect water outflow results as all of them are below than 10 s. However, it should be noted that based on microtexture and macrotexture, good results are coming from a brushing time in the range between 6 to 12 h.
- It should be mentioned that the results acquired from field investigation is faster than results acquired in laboratory tests. Where outflow times are more approximately 60 s and less than 10 s for laboratory tests and field investigation results,

respectively. This is occurred because the levelling process of ECC in full-scale application is considerably difficult, resulting in a rougher surface.

- Grooves on ECC-Cor specimens reduce the WFT on the surface. With the same total area (2 m × 0.5 m), the WFT on grooved ECC-Cor surface will be more than 50% decreased in comparison to ECC-Cor surface. The number, however, might vary on the dimension of the grooves (width, length, and thickness). In consequence, thinner WFT will cause a higher safety for motorists/vehicles as it creates smaller risks of the hydroplaning effect between the tire of vehicles and the pavement in wet condition.
- The validated 3D Computational Fluid (CFX) model gives a strong correlation with a lab test and equation results. In comparison to lab tests, the 3D CFX model gives 1.1 and 5.6% differences for 1 and 5% grades, respectively. While in comparison to (Eq. 33, the 3D CFX model gives 3.5 and 5.6% differences for 1 and 5% grades, respectively). This means 3D CFX model can be used to study and to design a surface water drainage. Several parameters can be studied with this model, including the effect of macrotexture, the effect of grade, and the effect of grooves to water film thickness.
- The roughness of the material, which is sand grain roughness in the CFX model, negatively affects the velocity of the water flow. The higher the roughness of the material applied, the slower the water flow is in the results.

Recommendations for further research are given:

- Numerical studies from different aspects can be done including the wet tire pavement interaction and hydroplaning model. Similarly, those numerical studies were done on different materials, tire, and scales.
- The OFT method can be used as one alternative for a quick assessment to evaluate the drainage performance based on texture properties in Singapore.

16.4 The Study of Correlation Between Skid Resistance and Surface Water Drainage Performance

Key contributions and findings from the study include:

- Due to the absence of coarse aggregates in ECC mix design, the skid resistance and surface water drainage performance of ECC is a concern for pavement applications. Corundum as fine aggregates was incorporated into ECC mix to improve the microtextures and skid resistance of ECC. The tensile strain capacity generally reduces with increasing corundum content. However, the resulting ECC-Cor still possesses a tensile strain capacity around 1% and flexural strength of 10 MPa, which are satisfactory for most pavement applications.
- Addition of corundum significantly improves the microtexture of ECC, resulting high a BPN more than 70. ECC-Cor shows good durability and the microtextures can last for more than 100,000 load cycles.

- ECC-Cor Spread method can be a good alternative to fulfil skid resistance requirements for pavement application. This method maintains the mechanical properties of normal ECC, while the brushed surface of corundum layer acts as a surface layer to provide the skid resistance.
- While the addition of corundum enhances the microtexture and skid resistance of ECC, improvement on macrotexture and surface water drainage is marginal and the risk of hydroplaning still exist. Further surface treatment with grooves successfully improves the functional performance, not only by minimizing the risk of hydroplaning (OFT < 4 s) but also by improving the skid resistance (BPN > 90).
- It was found that pavement geometries affect surface water drainage performance, particularly for surfaces with low MTD. The empirical model proposed in the current study is able to correlate the relation between water drainage (OFT) and MTD and pavement geometries.

Recommendations for further research are given:

- With the inclusion of synthetic fibers with high aspect ratios, practical use of ECC-Cor for full scale pavement applications should ensure few aspects. Properties of ECC-Cor produced in large scale processing should be maintained by selecting suitable raw materials, mix design, mixer, and mixing sequence and by controlling fresh properties of paste to ensure uniform fiber dispersion.
- With high unit volume cost of ECC-Cor, it is advisable to use the material strategically. For example, instead of using ECC-Cor to build full depth pavement, the overlay approach greatly reduces the required volume of material.
- ECC-Cor pavement can be an economical solution from the life cycle cost perspective by reducing rehabilitation and maintenance frequency and extending service life. Further study is necessary for large scale processing of ECC-Cor and life cycle and life cycle cost assessments of ECC-Cor pavement.

16.5 The Study of Noise

Key contributions and findings from the study include:

- The evaluation of the Mean Profile Depth (MPD)-value, according to (ISO 13473-1 2019) [3], and the shape factor of the profile give a good indication whether the road surface is suitable to reduce the tire-road noises.
- The noise level coming from the pavement made of ECC can be up to 65 dB(A), which is quite safe as it is below of many standard values regulated by the authorities. However, from the perspective of skid resistance and surface water drainage performance, the ECC slab is not safe to be used for roadways.
- The noise level for ECC-Cor-Gro 1 increased up to 69.2 dB(A) because of the lower shape factor (G) due to a rougher surface as results from the existence of corundum aggregate on the ECC surface. The shape factor at ECC-Cor is at 55% which is very low for dense surfaces.

- It can be seen from the analysis from four parameters namely vibration, air, cavity, and residual, that the biggest differences can be found in the vibrational noises, meaning that ECC-Cor has a rough surface that stimulates the tire to vibrate more.
- The modification of ECC-Cor by introducing grooves (ECC-Cor-Gro 1) has successfully reduced the sound level coming from the vibration. The sound level from the air parameter is decreased slightly, mainly in the frequency from 315 to 1250 Hz. The shape factor is then increased for ECC-Cor-Gro 1.
- The ECC-Cor with grooves can be a solution for the (electrified or normal) roadways with pavement made of ECC in Singapore condition, where the National Environment Agency (NEA) of Singapore regulates the noise limit for 67 dB to new roads and expressways [2].

Recommendations for further research are given:

- As has been discussed, three parameters significantly affect noise levels for pavement namely stiffness, porosity, and texture. For further research, ECC can be modified to decrease the stiffness and or increase the porosity. However, the study needs to be comprehensive as changing those two parameters will affect the mechanical properties and durability performance.
- Absorption noise test can be done to see how the noise level is affected by the material itself. The test is more suitable for experimental test in relation to changing the density and the porosity of ECC material.
- Field test noise related on full scale pavement application made of ECC can be done to see direct results when there is a minimum length of pavement made of ECC applied on real roadways.

16.6 The Study of Self-cleaning and Slip-proof

Key contributions and findings from the study include:

- Different T_iO_2 nano-particles were tested first to evaluate their photocatalysis property under visible light irradiation. Among the four commercially available T_iO_2 nano-particles, KRONOClean K7000 possessed the best photocatalytic performance under visible light irradiation due to high surface area and porosity, providing more active sites for reaction.
- Experiments were carried out to study the effects of white silica sand content and T_iO_2 dosage on the mechanical properties, micro- and macrotexture for skid resistance related properties, self-cleaning performance, and dirt pick-up resistance of the resulting white ECC. Mechanical tests showed that the compressive strength and skid resistance increase while the tensile strength and tensile strain capacity of white ECC decrease with increasing silica sand content. Furthermore, compressive strength, tensile strength, tensile strain capacity, self-cleaning performance, and dirt pick-up resistance of white ECC increase with increasing T_iO_2 dosage.
- The inclusion of T_iO_2 in ECC-SS successfully engages photocatalytic function under the visible-light, facilitates the decomposition of RhB. The addition of T_iO_2 increases self-cleaning performance.

- X-ray photoelectron spectroscopy (XPS) analysis revealed that K7000 is a type of carbon modified T_iO_2 which might improve charge separation and electron transfer process, leading to improved performance under visible light/low light condition.
- KRONOClean K7000 was then selected for the development of the bright and slip-proof ECC with the inclusion of white silica sand. With 20% silica sand and 5% T_iO_2 inclusion, the resulting white ECC possess excellent mechanical properties (i.e., compressive strength >50 MPa, tensile strength >4.5 MPa, and tensile strain capacity >4%) and high skid resistance (~80 BPN) and has high surface reflectivity with a brightness value of 90.
- The resulting white ECCs shows strong visible light activated photocatalysis property (i.e., 90% reflectance recovery after 4 h visible light irradiation, 8% surface brightness reduction after DPUR test).

Recommendations for further research are given:

- With the inclusion of synthetic fibers with high aspect ratios, practical use of ECC-SS-T_iO_2 for full scale pavement applications should ensure the properties produced in large scale processing. It should be maintained by selecting suitable raw materials, mix design, mixer, and mixing sequence and by controlling fresh properties of paste to ensure uniform fiber dispersion.
- Further study is necessary for the durability of ECC-Cor/ECC-SS with T_iO_2 and not only for the durability of the surface texture, but also for the self-cleaning performance.
- Further study on life cycle cost assessments of ECC-Cor/ECC-SS with T_iO_2. for pavement application is needed.

References

1. Land Transport Authority (2010) Materials and workmanship specification for civil and structural works. Land Transport Authority, Singapore
2. Bhanap I (2013) An analysis of roadway noise at residential estates in close proximity to expressways in Singapore. Noise Health 15(64):183–189
3. ISO 13473-1 (2019) Characterization of pavement texture by use of surface profiles

Appendix A

Dynamic Power Transfer Components, PUTW Design and Construction Method, and ECC Material for Electrified Roadway

The electrified roadway concept was introduced decades ago. One of the first electrified roadways projects using Dynamic Power Transfer is done by the California Partners for Advanced Transit and Highways (PATH), Institute of Transportation Studies, UC Berkeley, and the Southern California Association of Governments (SCAG) in the 1980s. The project is called The Roadway Powered Electric Vehicle (RPEV). Within the project, an approximately 200 m road track was constructed. The track consisted of two 60 m electrified segments. An electric bus of 35 passengers with a maximum speed of 64 km/h was used [1]. The details of the RPEV components can be seen in Fig. A.1, which illustrates the design and the actual components. The RPEV component details are summarized in Table A.1.

Another innovative power supply for dynamic power transfers, OLEV, was developed by KAIST [2]. OLEV is developed by using a thin rail as a module. The first successful demonstration, identified as the first generation (1G), was done in 2009. In about 10 years, the innovation has made significant progress. They manage to reduce the rail width and to increase the output power with relatively high efficiency [3]. The development progress is summarized in Table A.2. The latest developments, S-Type, has been successfully reducing the rail width to only four cm wide. This updated design makes it easier to fold itself by using flexible power cables (Fig. A.2).

The development of modified ECC material used in this study is based on micromechanics-based design, known as the Performance-Driven Design Approach (PDDA). As shown in Fig. A.3, this approach emphasizes micromechanics models as the unifying bridge between mechanical properties and material microstructure. It imposes requirements of structural performance, e.g., deformation capacity, ductility, and load capacity, that are depicted into mechanical properties at the material level, e.g., strength, ultimate strain, toughness, and maximum crack width.

A. A. Bawono, *Engineered Cementitious Composites for Electrified Roadway in Megacities*, Springer Theses, https://doi.org/10.1007/978-3-030-88542-7

Table A.1 RPEV system dimensions

System	Description	Units
Inductive coupling system	Airgap height	6 cm
	On board controller weight	395 kg
Pickup	Length	436 cm
	Width	100 cm
	Weight	545 kg
	Core length	15 cm
	Core width	50 cm
	Total pickup core weight	354 kg
Roadway inductor core modules	Width	51 cm
	Length	280 cm
	Thickness	11 cm
	Weight	409 kg
Roadway conductors	Aluminium cable	1.8 cm diameter
	Aluminium busbar	0.64 × 3.81 cm

Table A.2 OLEV development

	1G (Bus)	2G (Bus)	3G (Bus)	4G (Bus)	I-type	S-type
Rail width	140 cm	80 cm	80 cm	80 cm	10 cm	4 cm
Air gap	17 cm	23 cm	23 cm	23 cm	20 cm	20 cm
Output power (pick up unit)	12 kW	15 kW	33.3 kW	20 kW	27 kW	22 kW
EMF	<51 mG	<50 mG	<62.5 mG	<62.5 mG	<10 mG	<10 mG
Efficiency	72%	81%	85%	85%	74%	71%

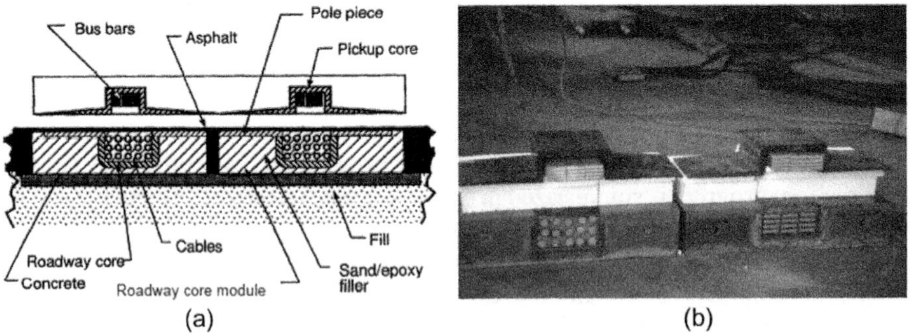

Fig. A.1 Onboard power inductors for RPEV **a** real **b** illustration

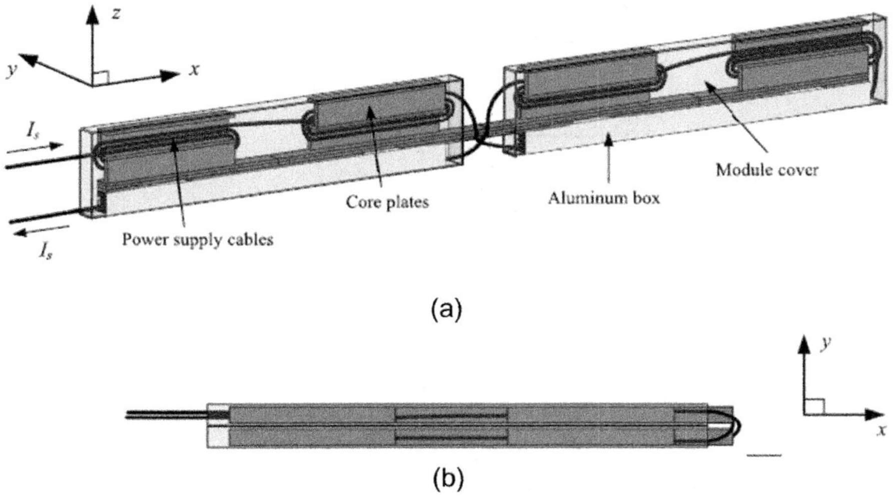

Fig. A.2 Configuration of the proposed ultra slim S-type power supply modules including two magnetic poles. **a** Bird's eye view for two unfolded modules. **b** Top view of a folded module

This approach offers a systematic engineering solution rather than trial-and-error empirical experiments [4].

The Precast Ultra-Thin Whitetopping (PUTW) slab can be designed in various dimensions, depending on the traffic load, temperature load, and supporting structural layers. In this study, the PUTW slab was manufactured with a size of 3400 mm × 2400 mm with a thickness of 70 mm. The slab built with grouting ports and lifting/levelling devices embedded. The design of the PUTW slab can be seen in Fig. A.4.

The PUTW slab is built based on a prefabricated method. The manufacture of the PUTW slab consists of several steps, including preparation of the steel formwork, placement of silicone mat to form the convex nodes for shear, embedment of the lifting/levelling device (details of lifting/levelling device can be seen in Fig. A.5), and grouting port. Only then, the casting process with ECC material can start.

The casting is done using two different materials in two steps, ECC and modified ECC with corundum aggregates. Firstly, the ECC material is poured into the steel formwork up to 60 mm thickness. After around 15 min, the second layer, modified ECC with corundum aggregates, is poured above the first layer. Both materials are still in fresh condition, and they are not completely hardened. Thus, both materials will be bonded soundly to each other. After 18–36 h, the surface needs to be brushed to expose the corundum aggregates. The manufacturing process of the PUTW slab is illustrated in Fig. A.6.

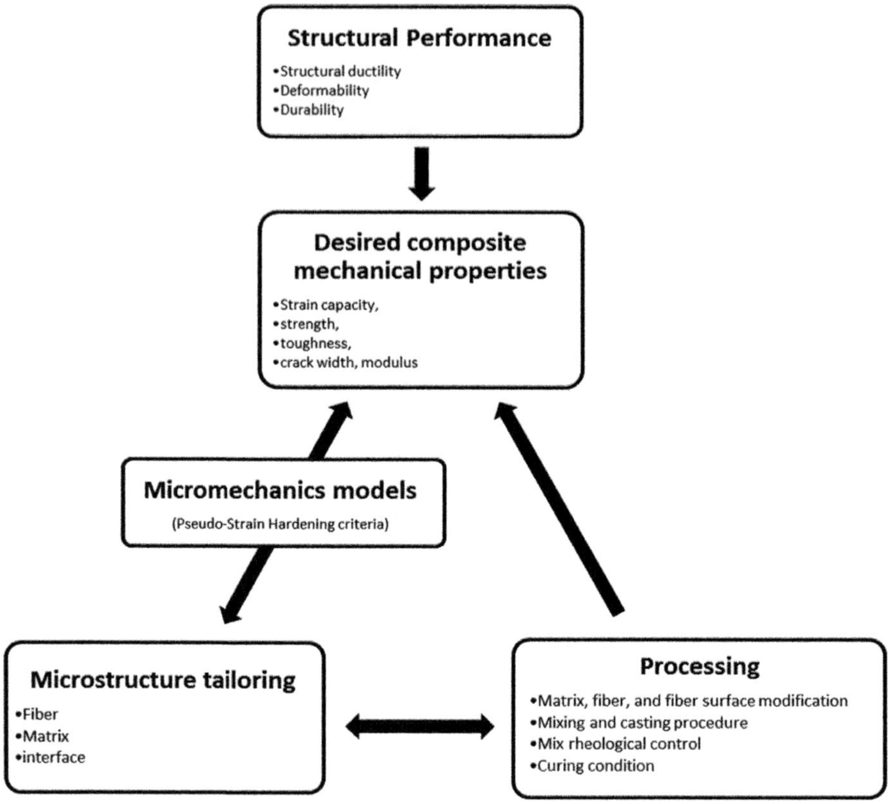

Fig. A.3 PDDA approach on ECC design [4]

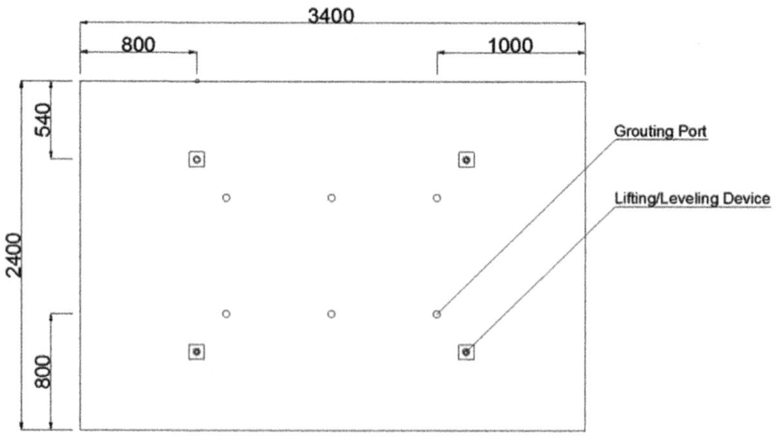

Fig. A.4 Dimension of PUTW slab and location of grouting port and lifting/leveling device

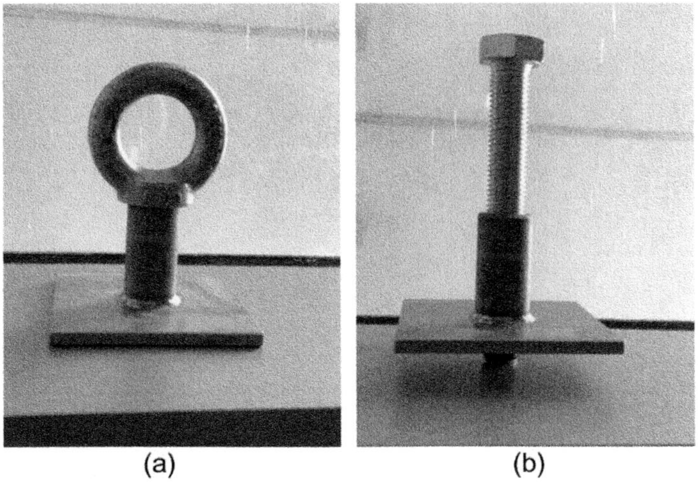

Fig. A.5 Lifting/leveling device **a** lifting bolt. **b** levelling bolt

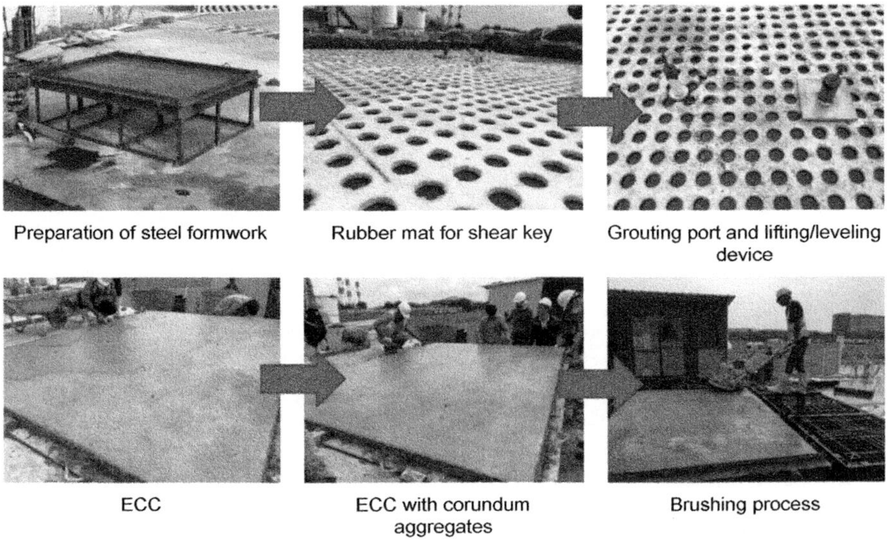

Fig. A.6 Steps of PUTW manufacture process

References

1. Shladover S (2006) PATH at 20—history and major milestones
2. Choi SY, Jeong SY, Gu BW, Lim GC, Rim CT (2015) Ultraslim S-type power supply rails for roadway-powered electric vehicles. IEEE Trans Power Electron 30(11):6456–6468. https://doi.org/10.1109/TPEL.2015.2444894

3. Suh NP (2017) The on-line electric vehicle: wireless electric ground transportation systems. Springer, Cham
4. Li VC, Wang S, Wu C (2001) Tensile strain-hardening behavior of polyvinyl alcohol engineered cementitious composite (PVA-ECC). ACI Mater J 98(6):483–492

Appendix B

Consideration of Functional Performance Analysis in Typical Construction Process

An approach of novel comprehensive functional performance analysis is proposed in this study. The approach can be used for new construction and retexturing (rehabilitation construction) projects. Typically, a construction project consists of a few steps, starting from defining the objective of the project, collecting data, determining the design method, selecting the preferred construction method, and then doing the construction. The improved analysis steps are shown as a flow chart in Fig. B.1.

The first step in a typical construction project is to define the target objective of the project. For example, the authority (decision-makers) can set a project target to improve the skid resistance, reduce the traffic noise, improve the surface water

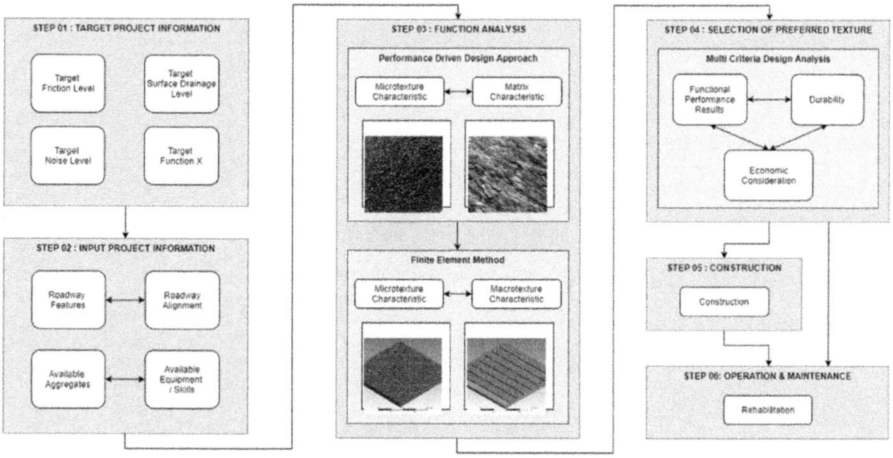

Fig. B.1 Flow chart for comprehensive analysis

A. A. Bawono, *Engineered Cementitious Composites for Electrified Roadway in Megacities*, Springer Theses, https://doi.org/10.1007/978-3-030-88542-7

drainage, and reduce emission on roadways. Many times, the objective of the project can also be more than one objective. For example, road construction needs to fulfil high skid resistance and surface water drainage requirements.

Secondly, once the target of the project is defined, the designer needs to collect all related data information as an input for design. Data can include roadway features, roadway alignment, availability of aggregates near to the site, availability of equipment, and availability of contractors/sub-contractors for the project.

As for the third step, a comprehensive analysis proposed in this study, specific to designing the functional performance of the roadway, can be used. The Performance-Driven Design Approach (PDDA) [1] can be modified and used to design the optimum material and matrix design based on their structural and functional performance. Further, numerical models can be utilized to design and predict the performance of the pavement. A good validation with laboratory and or field test is necessary.

Many times, multiple outputs of design coming from the numerical analysis are observed. The Multi-Criteria Design Analysis (MCDA) can be utilized in attempting to identify the most optimum design [2]. Authority (decision-makers) can use the MCDA, which can contain a few parameters, including functional performance, durability, and economical, for careful and objective consideration. Finally, once the design is decided, the construction can start.

References

1. Li VC, Wang S, Wu C (2001) Tensile strain-hardening behavior of polyvinyl alcohol engineered cementitious composite (PVA-ECC). ACI Mater J 98(6):483–492
2. Ashby MF (1999) Materials selection in mechanical design. Butterworth-Heinemann, Oxford

Appendix C

Multi-criteria Design Analysis on Functional Performance

In many cases, a design (including pavements) requires coupled optimization of two or more parameters of performance. An example, given by Ashby [1], a material was needed for a beam for a bicycle component. The objective is to minimize the weight as well as the cost. The designer wants to have light material, but the marketing manager requires it to be as cheap as possible. Therefore, the material has a constraint of stiffness and price. In this study, MCDA is proposed to identify criteria including safety and other important parameters. Those criteria then are being scored and weighted. The weight proportion used in this study is based on a study done by NCHRP [2]. It was summarized that safety has more prioritization than other criteria (e.g., comfort).

The first analysis is done with the radar chart to evaluate each texturing method. A radar chart, also known as spider chart, web chart, polar chart, or star plot, is a two-dimensional chart to display multi-dimensional data structure. This spider chart is useful as it can be used to compare the average score of different characteristics of constraints (skid resistance, surface water drainage, tire-pavement noise, pavement roughness, durability, and cost). The functional performance of each texturing method for asphalt and concrete pavements are shown in Figs. C.1 and C.2, respectively. Seven constraints are considered for each texturing method, including microtexture, macrotexture, pavement roughness, surface water drainage, tire-pavement noise, durability, and cost. They are rated on a scale from 0 to 5.

The second analysis is done by plotting a two-constraints scenario. This analysis allows a comparison of many texturing methods. However, it is limited to two constraints. The plot is not determined to give a correlation between constraint (criterion). Instead, it gives a picture of what is the performance of two constraints for each texturing method. The results are shown in Fig. C.3. The figure is using a code to represent each texturing method. For asphalt pavement (coded as AC), there are about 35 different texturing methods found. Those texturing methods are then labelled with

A. A. Bawono, *Engineered Cementitious Composites for Electrified Roadway in Megacities*, Springer Theses, https://doi.org/10.1007/978-3-030-88542-7

a code, as it will make it easier to be coded and to be visualized in charts later. On the other side, there are about 37 different texturing methods summarized for concrete pavements (coded as PCC). The texturing method can be categorized into two: New and or overlay (coded as NO), and retexturing or friction restoration (coded as RF). Examples of the meaning of the coding are given: ACNO means texturing method for Asphalt Pavement for New and Overlay. Meanwhile, PCCRF means a texturing method for Concrete pavement for Retexturing and or Friction restoration. Details of code and name representation are summarized in Table C.1.

Figure C.3a shows a plot chart of microtexture against macrotexture for different texturing methods. There are various texturing methods available, resulting in good skid resistance and macrotexture for asphalt and concrete pavements. It is noted from the plot that diamond grooving is giving one of the most optimum design in terms of microtexture and macrotexture. Chip seal and longitudinal diamond grooving provide the second-best optimum design.

Figure C.3b gives an overview of texturing methods regarding microtexture and surface water drainage constraints. The best scenario is given by diamond grooving for concrete pavement and rubberized open-graded asphalt concrete. A surface with a water outflow of fewer than ten seconds results in good surface drainage resulting in lower hydroplaning risks.

Figure C.3c shows texturing methods performance regarding microtexture and tire-pavement noise. The first best option is given by heavy turf drag and longitudinal tining. The second-best options include diamond grooving and rubberized open-graded asphalt concrete.

Figure C.3d shows texturing methods performance regarding microtexture and pavement roughness. The first best options include longitudinal heavy turf drag, diamond grooving, and Nova chip. The second-best options include longitudinal diamond grooving, rubberized open-graded asphalt concrete, and asphaltic concrete friction course.

Figure C.3e shows texturing methods performance regarding macrotexture and surface water drainage. The first best options include whitetopping with tining and shot abrading. The second best include longitudinal diamond grooving.

Figure C.3f shows texturing methods performance regarding macrotexture and tire-pavement noise. The first best option includes an open-graded asphalt friction course. The second-best options include longitudinal diamond grooving, and longitudinal tining.

Figure C.3g shows texturing methods performance regarding macrotexture and pavement roughness. The first best options include longitudinal turf drag, astroturf drag, and Nova chip. The second-best options include transverse tining, rubber modified asphalt course, and longitudinal diamond grooving.

Figure C.3h shows texturing methods performance regarding surface water drainage and tire-pavement noise. The first best options include asphalt rubber friction course, longitudinal turf drag, and rubberized open-graded asphalt concrete. The second-best options include Porous European Mix.

Figure C.3i shows texturing methods performance regarding surface water drainage and pavement roughness. The first best options include rubberized bonded

Table C.1 Code and name of texturing method for asphalt pavement

Name of texturing method	Name code	Application code	
Asphaltic concrete friction course (ACFC)	A01	ACNO	
Asphalt rubber friction course (ARFC)	A02	ACNO	
Dense coarse graded asphalt concrete	A03	ACNO	ACRF
Open graded asphalt friction course (OGFC)	A04	ACNO	ACRF
Permeable friction course (PFC)	A05	ACNO	
Porous graded friction course (PGFC)	A06	ACNO	
Polymer modified HMA	A07	ACNO	ACRF
Porous European mix (PEM)	A08	ACNO	
Rubber modified asphalt concrete (RUMAC)	A09	ACNO	
Rubber modified HMA	A10	ACNO	ACRF
Rubberized asphalt concrete (RAC)	A11	ACNO	
Rubberized asphalt concrete (RAC) crumb rubber	A12	ACNO	
Rubberized open asphalt concrete (RAC-O)	A13	ACNO	
Stone mastic asphalt	A14	ACNO	
Superpave	A15	ACNO	
Rubberized open graded asphalt concrete (ROGFC)	A16	ACNO	
HMA ultra thin bonded wearing course	A17	ACNO	ACRF
Rubberized bonded wearing course (RBWC)	A18	ACNO	
Semi rigid	A19	ACNO	ACRF
Taconite HMA	A21	ACNO	ACRF
Bonded wearing course (BWC)	A22	ACNO	
ASPEN sealer	A23	ACRF	
Cold mix SMA overlay	A24	ACRF	
Carbonyte sealer	A25	ACRF	
Micromilling	A26	ACRF	
Slurry seal	A27	ACRF	
Polymer modified slurry seal	A28	ACRF	
Chip seal	A29	ACRF	
Ital grip	A30	ACRF	
Fog seal	A31	ACRF	
Microsurfacing	A32	ACRF	
Nova chip	A33	ACRF	
Cargill SafeLane*	A34	ACRF	
Skidabrader	A35	ACRF	ACRF

wearing course, rubberized open-graded asphalt concrete, transverse tining, longitudinal tining. The second best include Porous European Mix, and longitudinal turf drag.

Figure C.3j shows texturing methods performance regarding tire-pavement noise and pavement roughness. The first best options include rubberized bonded wearing course, asphalt rubber friction course. The second-best options include diamond grooving, Nova chip, rubberized open-graded asphalt concrete polymer-modified HMA. As expected, texturing methods for asphalt pavement able to generate less noise compared to those for concrete pavement generally (Table C.2).

References

1. Ashby MF (1999) Materials selection in mechanical design. Butterworth-Heinemann, Oxford
2. Hall J, Smith K, Littleton P (2009) Texturing of concrete pavements: NCHRP Report 634, Washington, D.C.

Table C.2 Code and name of texturing method for concrete pavement

Name of texturing method	Code	Application	
Carpet drag longitudinal	P1	PCCNO	
Broom drag longitudinal	P2	PCCNO	
Broom drag longitudinal diamond grooving	P3	PCCNO	
Turf drag longitudinal	P4	PCCNO	
Turf drag longitudinal skidabrader	P5	PCCNO	
Heavy turf drag longitudinal	P6	PCCNO	
Astroturf drag longitudinal	P7	PCCNO	
Astroturf drag longitudinal tining longitudinal	P8	PCCNO	
Burlap drag longitudinal	P9	PCCNO	
Tining transverse	P10	PCCNO	
Tining longitudinal	P11	PCCNO	
Tining transverse burlap drag	P12	PCCNO	
Tining longitudinal burlap drag	P13	PCCNO	
Tining transverse turf drag	P14	PCCNO	
Tining longitudinal turf drag	P15	PCCNO	
Tining transverse random	P16	PCCNO	
Tining longitudinal random	P17	PCCNO	
Tining skew	P18	PCCNO	
Diamond grinding longitudinal	P19	PCCNO	PCCRF
Diamond grinding transverse	P20	PCCNO	PCCRF
Diamond grinding skew	P21	PCCNO	PCCRF
Diamond grooving longitudinal	P22	PCCNO	PCCRF
Diamond grooving transverse	P23	PCCNO	PCCRF
Diamond grooving skew	P24	PCCNO	PCCRF
Diamond grooving longitudinal burlap drag	P25	PCCNO	PCCRF
Diamond grooving transverse broom drag	P26	PCCNO	PCCRF
Diamond grooving longitudinal turf drag	P27	PCCNO	PCCRF
Diamond grooving transverse turf drag	P28	PCCNO	PCCRF
Porous PCC	P29	PCCNO	
Exposed aggregate concrete (EAC)	P30	PCCNO	
Whitetopping tining random	P31	PCCNO	PCCRF
Ultra-thin whitetopping (UTW)	P32	PCCNO	PCCRF
Thin whitetopping (TTW)	P33	PCCNO	PCCRF
Next generation concrete surface (NGCS)	P34	PCCNO	
Shot abrading	P35	PCCRF	
VDOT modified EP-5	P36	PCCRF	
Smooth troweled	P37	PCCRF	

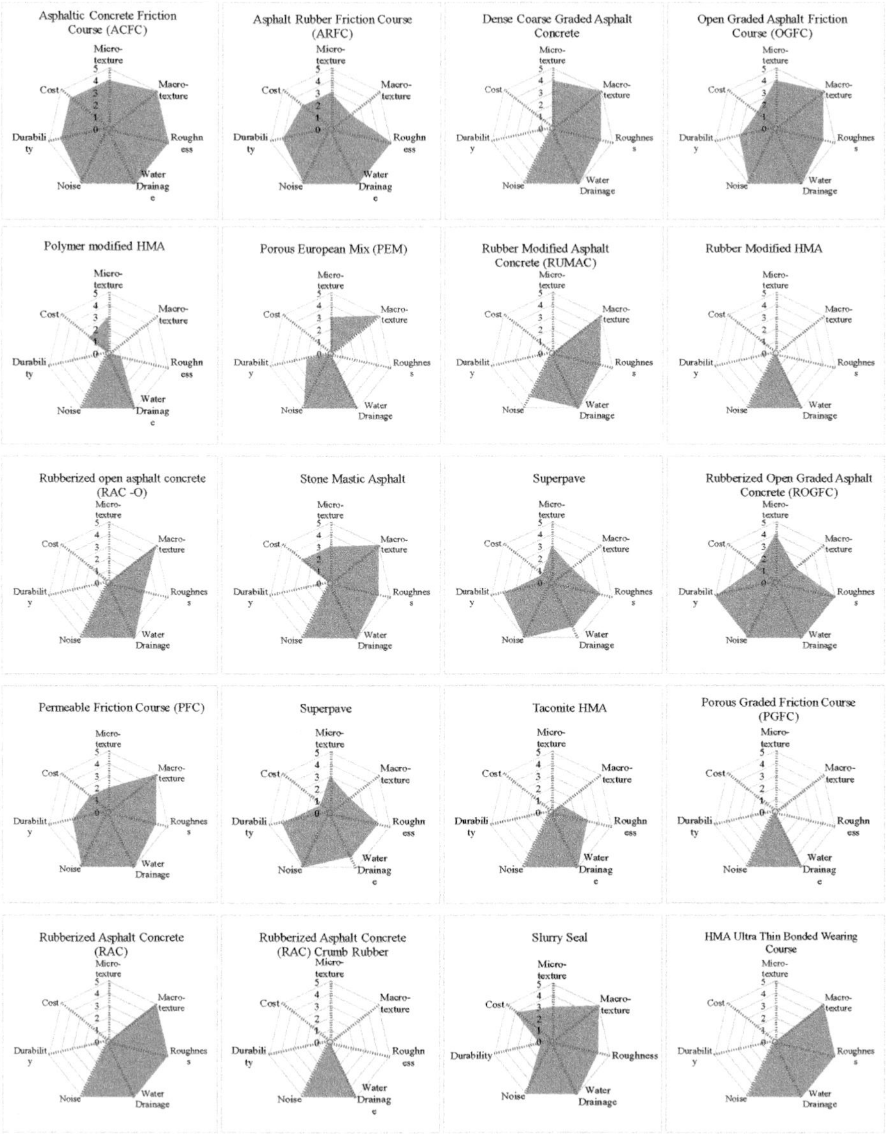

Fig. C.1 Functional performance on asphalt pavement texturing methods

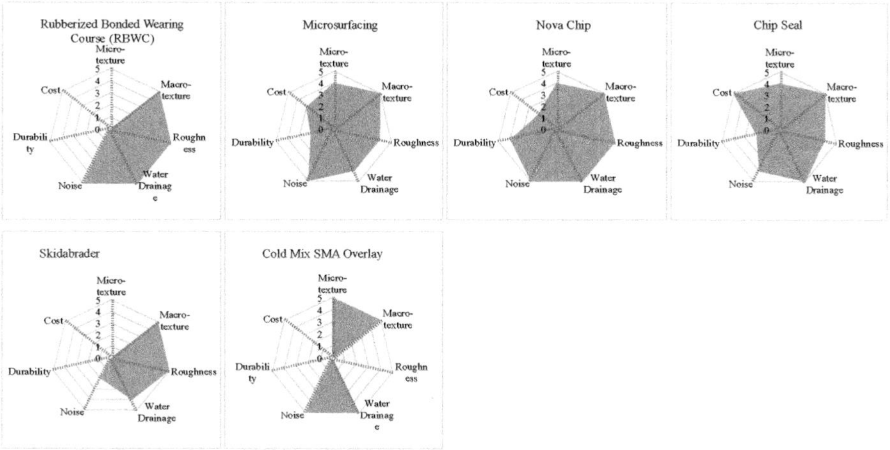

Fig. C.1 (continued)

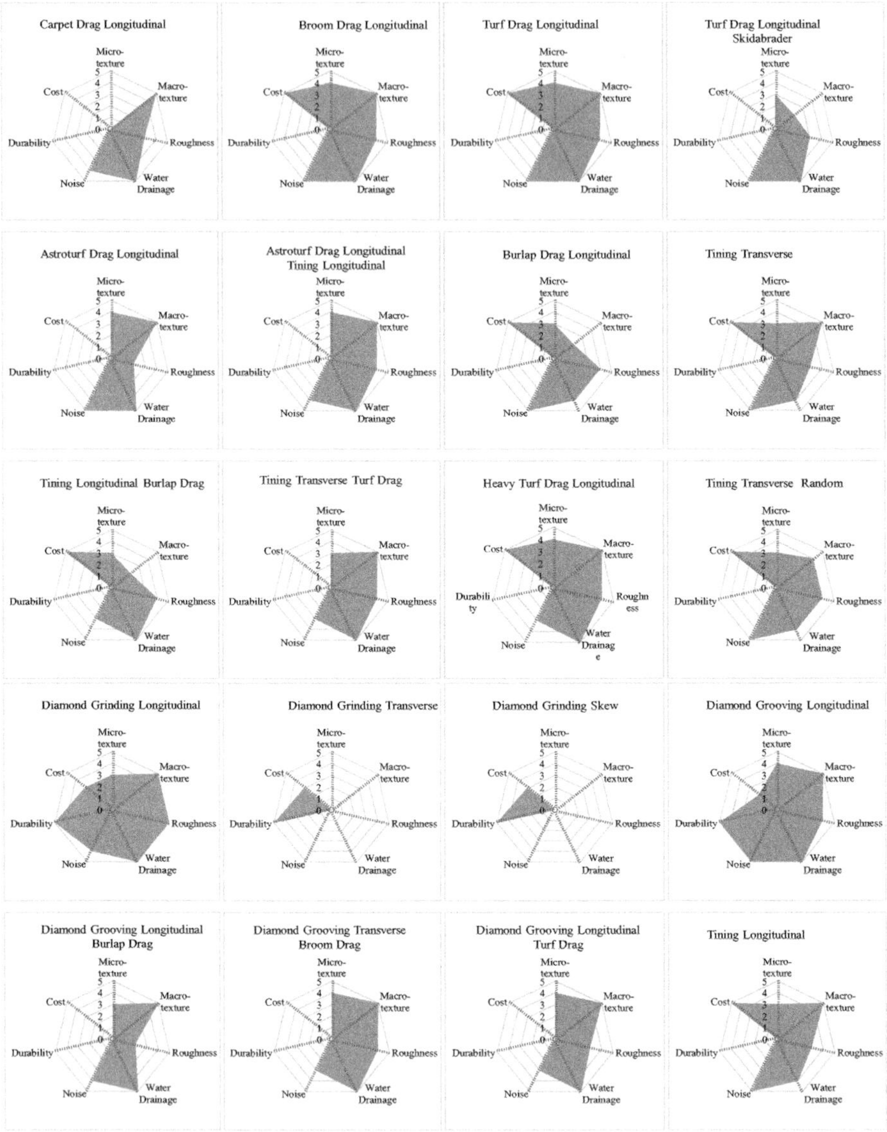

Fig. C.2 Functional performance based on concrete pavement texturing methods

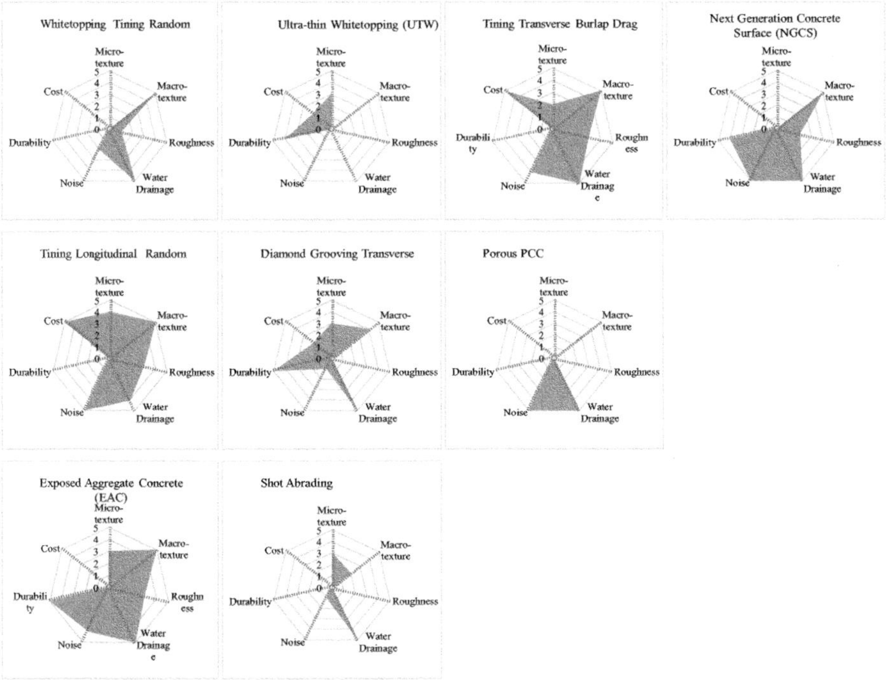

Fig. C.2 (continued)

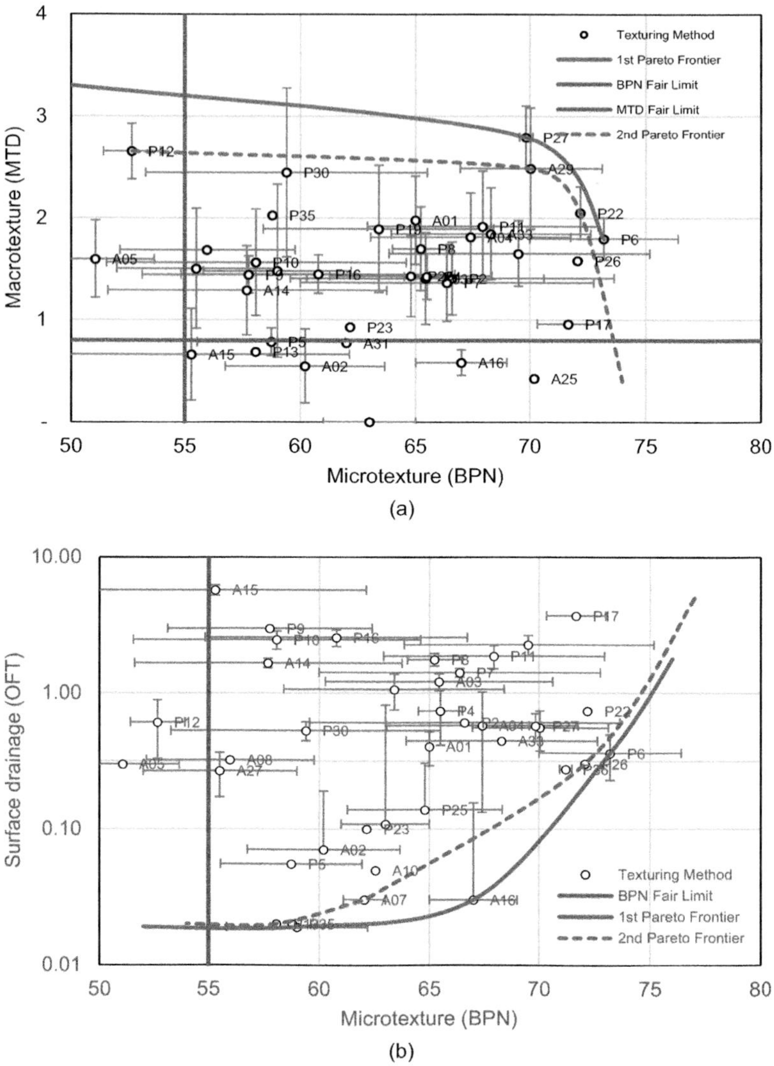

Fig. C.3 Two-constraints analysis of functional performance for pavement **a** microtexture and macrotexture, **b** microtexture and surface water drainage, **c** microtexture and tire-pavement noise, **d** microtexture and pavement roughness, **e** macrotexture and surface water drainage, **f** macrotexture and tire-pavement noise, **g** macrotexture and pavement roughness, **h** surface water drainage and tire-pavement noise, **i** surface water drainage and pavement roughness, and **j** tire-pavement noise and pavement roughness

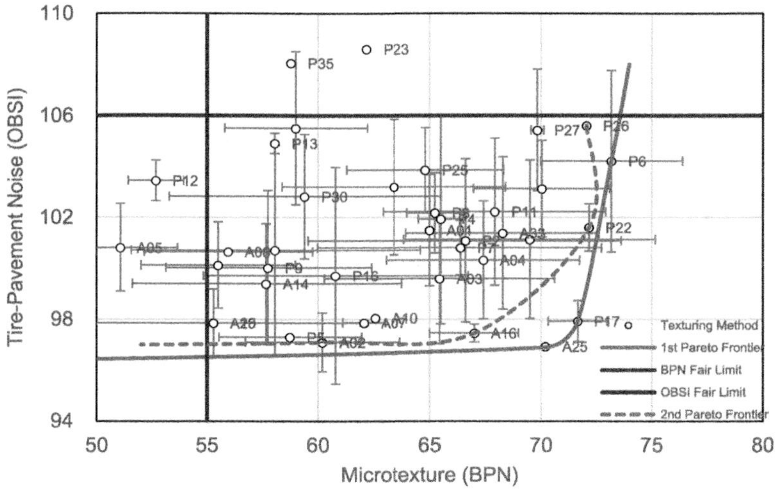

(c)

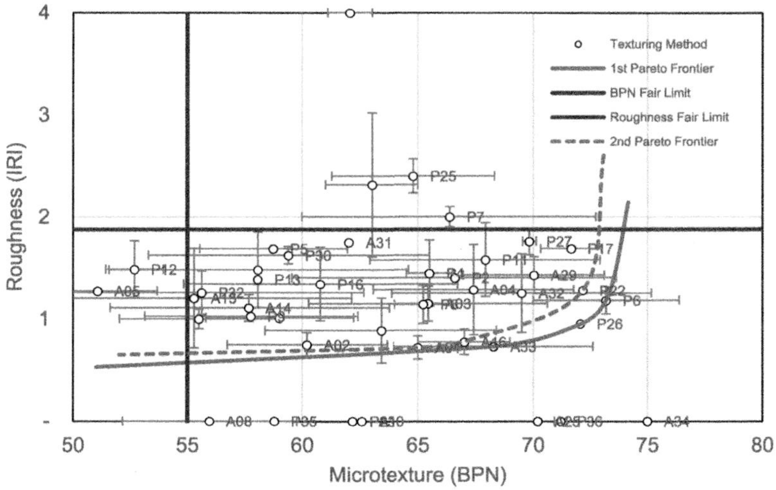

(d)

Fig. C.3 (continued)

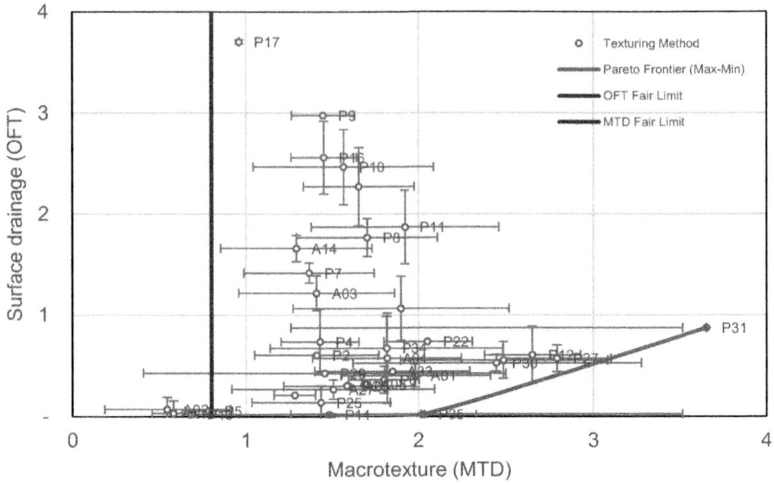

(e)

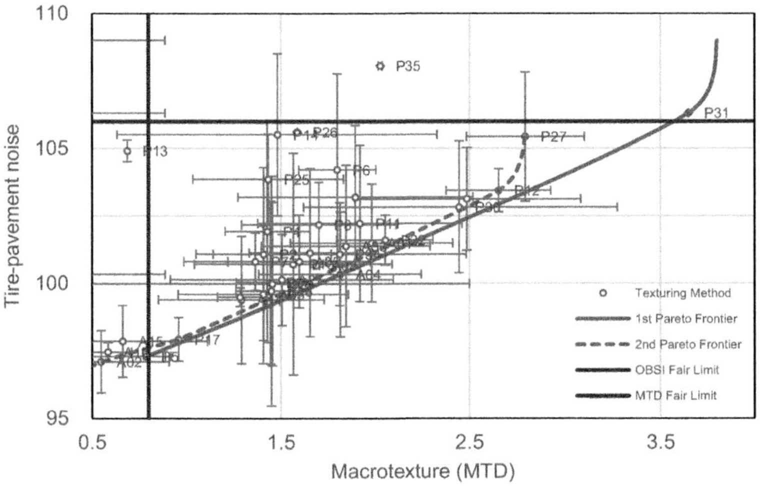

(f)

Fig. C.3 (continued)

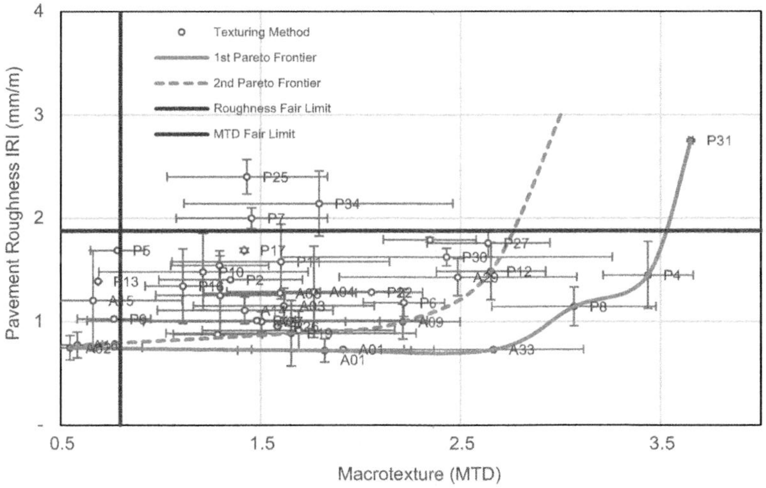

(g)

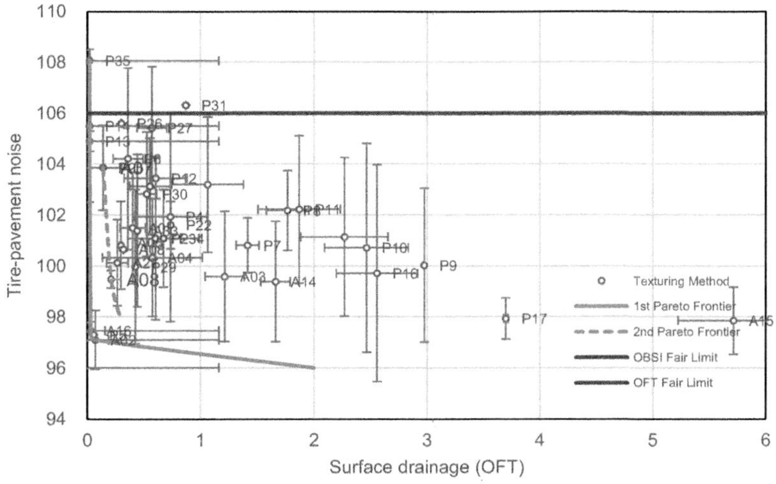

(h)

Fig. C.3 (continued)

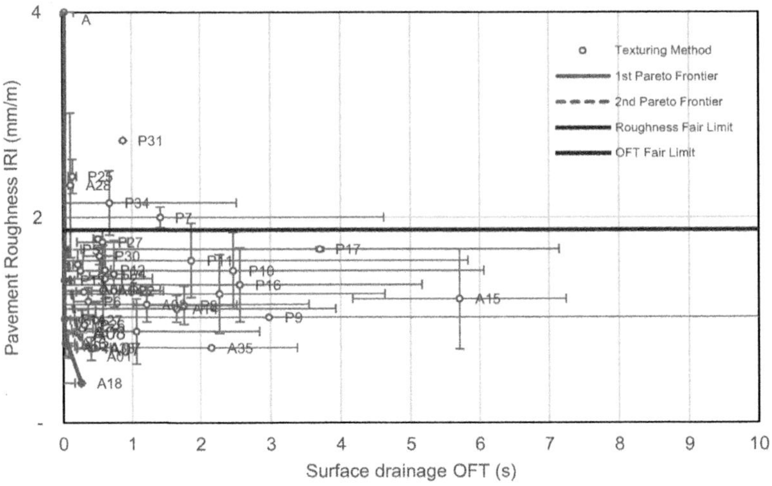

(i)

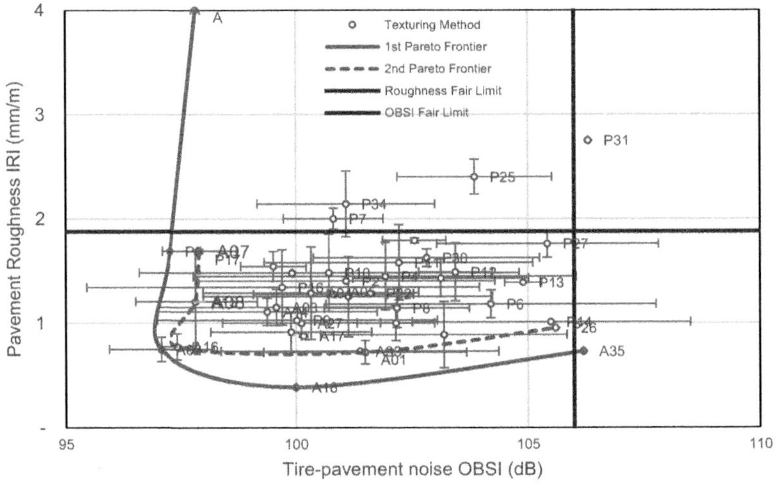

(j)

Fig. C.3 (continued)

Appendix D

Standards for Skid Resistance on Different Countries

Roadway authorities (decision-makers) are implementing high standards for their new and improved roads to enhance the motorist's safety. However, different countries or authorities apply a different standard.

In Singapore, the skid resistance standard is referring to the Code of Practice (COP) for Works on Public Streets [1] and Materials and Workmanship specification [2]. According to the COP, a concrete surface paving on footpath shall be approved sprayed-on acrylic modified UV stable pigmented cementitious topping with a minimum thickness of one mm, minimum cured compressive strength of 45 N/mm^2 and skid resistance not less than 50. Concrete Surface paving on the carriageway shall be approved spray-on acrylic modified UV stable pigmented cementitious topping with a minimum thickness of one mm, minimum cured compressive strength of 60 N/mm^2 and skid resistance not less than 65. While according to the specification, the minimum limit of skid resistance for asphaltic pavement is 55 (Table D.1).

In Germany, the skid resistance standard follows the German specification of ZTV Asphalt-StB 07 and ZTV Beton-StB 07. It stated that the friction coefficient for road pavements must never fall short of the limits by 0.03 with SCRIM measurement [3]. The friction coefficient is summarized in Table D.2.

The friction level for a new construction and surface restoration in Switzerland is measured using Skiddometer BV-8 or the Stuttgarter Reibungsmesser. As shown in Table D.3, the friction levels, which are defined based on the speed limit.

Table D.1 Standard of skid resistance in Singapore

Category	Site definition	Skid resistance (BPN)
Surface coating	Paving on carriageway	$\geq$65
Concrete surface	Paving on footpath	$\geq$50
Asphaltic	Open graded asphalt	$\geq$55

A. A. Bawono, *Engineered Cementitious Composites for Electrified Roadway in Megacities*, Springer Theses, https://doi.org/10.1007/978-3-030-88542-7

Table D.2 Standard of skid resistance in Germany

Speed of measurement (km/h)	Limits μ_{SCRIM}	
	Acceptance value (new road)	Up to termination of guarantee
40	0.56	0.49
60	0.51	0.45
80	0.46	0.40

Table D.3 Friction levels for construction in Switzerland

Speed limit (km/h)	Using skiddometer BV-8	
	Speed	μ
≤60	40	0.48
>60 and ≤100	60	0.39
>100	80	0.32

Table D.4 Standard of skid resistance by TRRL

Category	Type of site	Minimum BPN
A	Difficult sites such as: Roundabouts, Bends with radius less than 150 m on unrestricted roads, Gradients 1 in 20 or steeper of lengths greater than 100 m, Approaches to traffic lights on unrestricted roads	65
B	Motorways, trunk and class 1 roads and heavily trafficked roads in urban areas (carrying more than 2000 vehicles per day)	55
C	All other sites	45

Note For Category A and B sites where the speed of traffic is high (in excess of 95 km/h) an additional requirement is a minimum texture depth of 0–65 mm

TRRL used British Pendulum Number to determine its skid resistance [4]. They categorized three different categories of skid resistance based on the different type of site (Table D.4).

The skid resistance standard is found to be different throughout the states in the USA. Some of the Departments of Transport (DOT) is using Skid Number (SN). In contrast, the other DOT uses Friction Number (FN) or even Polished Stoned Value (PSV) to determine the Investigatory Level related to Skid Resistance. Different values and actions are used by different DOT [5] (Table D.5).

Fewer standards are found for the new road, which measured the thresholds by the macrotexture of the surface pavement. Hall et al. mentioned MTD should be achieved within a range of 0.80–1.50 mm for the new pavement with longitudinal grooves [6]. However, some authorities (e.g., United Kingdom, Australia, and Austria) implement both microtexture and macrotexture in their standards.

The UK attempts to provide an MTD of 1.5 mm (0.06 in.) for new pavements [7]. A 16–10 mm coarse aggregate with a 1.5 mm texture depth is required to provide adequate skidding resistance. The Department of Transport in Australia recommends

Table D.5 Standard of skid resistance in different state in USA

State	Limit	Explanation
California	Office of traffic safety notified when SN40 is below 30	Varies
Idaho	Range in numbers	At 35, section is listed. At 30, section is recommendable
Illinois	Less than FN 40 = 35	FN $\leq$ 30: contributing to wet-pavement accidents 31 $\leq$ FN $\leq$ 35: friction as primary factor FN $\geq$ 36: some condition other than pavement friction
Indiana	Minimum friction number (FN40) is 30	
Louisiana	SN equal to or over 35 for new construction Friction Rating 1 (all mixtures): PSV > 37 Friction Rating 2 (all mixtures): PSV = 35–37 Friction Rating 3 (all mixtures, except wearing courses with ADT > 7,000 vehicle/day): PSV = 30–34 Friction Rating 4 (all mixtures except wearing courses): PSV = 20–29	
Michigan	ADT < 100 veh/day/lane: no AWI requirement ADT $\geq$ 100 and < 500 veh/day/lane: AWI $\geq$ 220 ADT $\geq$ 500 veh/day/lane: AWI $\geq$ 260	Special attention is given to investigate surfaces with SN at 30 or below
Ohio	Below 30—poor 30–40—adequate, depending on traffic and geometry Over 40—satisfactory	Some correction likely where SN = 30
Tennessee	Above FN 35: acceptable FN 30–35: questionable FN below 30: unsatisfactory	
Vermont	Monitor are if SN is 40 or below; inform proper personnel of problem when SN is below 30	
Virginia	Based on bald tire FN40 minimum of approximately 20	

a texture depth of 0.40 mm should be achieved for freeways and other high-class facilities with free-flowing traffic conditions, and a texture depth of 0.60 is the minimum limit for highways and other major main roads [8]. In Austria, an 8 mm maximum size aggregate was used to achieve good low-noise pavement properties. An 8–4 mm sized aggregate with a 1.0 mm texture depth is recommended for lower speed roads (90 km/h) where the risk of aquaplaning is less than for high-speed roads [9]. Singapore only specifies a criterion for acceptance of open-graded asphalt based on a texture depth of a minimum of 1.20 mm.

In typical pavement management system, the investigatory levels can be set, representing a limit, above which the skid resistance is considered to be satisfactory. But, they need to be set at or below a level where the road is judged to require an investigation of the skid resistance requirements. Once roads have been identified requiring treatment to improve the skid resistance, signs warning is ideally erected. Then the intervention level is set as a threshold when a surface roadway needs to be improved.

The investigatory level of skid resistance which is set 50 km/h (SCRIM) for pavement is stated in Design Manual for Roads and Bridges for Skid Resistance stated by the Highways Agency [10]. The investigatory level values in the UK and Victoria (Australia) are summarized in Tables D.6 and D.7, respectively.

Intervention levels based on friction can be determined with different measurements. The values are different depends on the level of roadways [7]. The minimum friction levels for intervention action are summarized in Table D.8.

The texture can also define the intervention levels. The requirements are higher for motorways compared to primary roads. The intervention levels based on texture in a few countries are summarized in Table D.9.

Table D.6 Standard of skid resistance by UK Highway Agency

Site category and definition		Site investigatory level					
		0.3	0.35	0.4	0.45	0.5	0.55
A	Motorway		■				
B	Dual carriageway non-event		■				
C	Single carriageway non-event			■	■		
Q	Approaches to and across minor and major junctions, approaches to roundabouts				■	■	■
K	Approaches to pedestrian crossings and other high risk situations					■	■
R	Roundabout				■	■	
G1	Gradient 5–10% longer than 50m				■	■	
G2	Gradient >10% longer than 50m					■	■
S1	Bend radius <500 m—dual carriageway				■	■	
S2	Bend radius <500 m—single carriageway					■	■

Notes

Investigatory Levels are for the mean skidding resistance within the appropriate averaging length

Investigatory Levels for site categories A, B, and C are based on 100 m averaging lengths (50 m lengths for some Overseeing Organizations) or the length of the feature if it is shorter

Investigatory Levels and averaging lengths for site categories Q, K, G and S are based on the 50 m approach to the feature but this shall be extended when justified by local site characteristics

Investigatory Levels for site category R are based on 10 m lengths

Residual lengths less than 50% of a complete averaging length may be attached to the penultimate full averaging length, providing the site category is the same

As part of site investigation, individual values within each averaging length should be examined and the significance of any values which are substantially lower than the mean value assessed

In Victoria (Australia), the investigatory level for skid resistance is divided into six site categories, depending on the traffic volume [11]

Table D.7 Standard of skid resistance in Victoria, Australia (VicRoads)

Site category	Site description	Investigatory level (SFC 50)	
		Primary roads and secondary roads >2,500 vehicles per lane per day	Secondary roads <2,500 vehicles per lane per day
1	Traffic light-controlled intersections Pedestrian/school crossings Railway level crossings Roundabout approaches	0.55	0.50
2	Curves with radius $\leq$250 m Gradients $\geq$5% and $\geq$50 m long Freeway/highway on/off ramps	0.50	0.45
3	Intersections	0.45	0.40
4	Maneuver-free areas of undivided roads	0.40	0.35
5	Maneuver-free areas of divided roads	0.35	0.30
		Investigatory level (SFC20)	
6	Curves with radius $\leq$ 100 m	0.60	0.55
7	Roundabouts	0.55	0.50

Table D.8 Intervention levels for friction

Road Category	Motorways	Primary	Secondary
Denmark	Speed < 80 km/h; μ = 0.4; Speed > 80 km/h; μ = 0.5		
Hungary	SCRIM > 0.50	SCRIM > 0.40	SCRIM > 0.33
Japan	Friction > 0.25		
The Netherlands	DWW > 38	DWW > 38	
New South Wales	SCRIM > 0.30–0.55		
New Zealand	SCRIM > 055 om event sites, 35 for no-event sites		
Quebec	SCRIM > 0.7	SCRIM > 0.7	SCRIM > 0.55
South Australia	BPN > 45	BPN > 45	BPN > 45
Switzerland	Same as for construction and rehabilitation		
The UK	Same as investigatory levels		
Arizona (The USA)	34 (μm)	34 (μm)	34 (μm)
Idaho	SN40S > 30	SN40S > 30	SN40S > 30
Illinois	SN40R > 30	SN40R > 30	SN40R > 30
Kentucky	SN40R > 28	SN40R > 25	SN40R > 25
New York	SN40R > 32	SN40R > 32	SN40R > 32
South Carolina	SN40R > 41	SN40R > 37	SN40R > 37
Texas	SN40R > 30	SN40R > 26	SN40R > 22

Table D.9 Intervention levels based on texture

Road category	Motorways	Primary	Secondary
Hungary	RMS < 0.22 mm	RMS < 0.14 mm	RMS < 0.10 mm
New Zealand	MTD < 0.90 mm	MTD < 0.90 mm	MTD < 0.90 mm
Quebec	MTD < 0.60 mm	MTD < 0.60 mm	MTD < 0.60 mm
South Australia	0.4 < MTD < 0.8 mm	0.2 < MTD < 0.4 mm	0.2 < MTD < 0.4 mm

References

1. Land Transport Authority (2000) Standard details of road elements. Land Transport Authority, Singapore
2. Land Transport Authority (2010) Materials and workmanship specification for civil and structural works. Land Transport Authority, Singapore
3. Lechner B (2013) Braking and Suspension (Road), Lecture Script. Technische Universität München, Munich
4. Road Research Laboratory Ministry of Transport (1969) Instructions for using the portable skid resistance tester. Road Research Laboratory, Crowthorne, Berkshire
5. Corsello P (1993) Evaluation of surface friction guidelines for Washington State Highways. Olympia, Washington
6. Hall J, Smith K, Titus-Glover L, Wambold J, Yager T, Rado Z (2009) Guide for pavement friction. Washington, D.C.
7. Henry J (2000) Evaluation of pavement friction characteristics: a synthesis of highway practice. Washington, D.C.
8. Department for Transport, Energy & Infrastructure (2016) Road surface texture measurement recommended investigatory levels. Australia
9. Highways Agency (2006) Surfacing materials for new and maintenance construction: pavement design and maintenance, 7th edn. United Kingdom
10. Highways Agency (2004) Skid resistance: pavement design and maintenance, vol 7. Design Manual Roads and Bridges, United Kingdom
11. Victorian Road Transport Agency (2015) Measurement and Interpretation of Skid Resistance using SCRIM. Technical Note TN-110, Australia

Appendix E

Surface Texture Measurement Test Method and Texturing Method

There are a few types of measurement test method for the surface texture of roadway. Each test method has its advantages and disadvantages. Those test methods for pavement surface texture are summarized in Table E.1.

There are various methods to improve the skid resistance of pavement for both asphalt concrete pavement. There are two different levels of improving skid resistance. The first level is texturing for new construction. The second level is re-texturing for maintenance or rehabilitation. Table E.2 summarizes the texturing techniques. The table is adopted and modified from NCHRP report number 108 [1].

References

1. Hall J, Smith K, Titus-Glover L, Wambold J, Yager T, Rado Z (2009) Guide for pavement friction. Washington, D.C.
2. Rasmussen RO, Bernhard R, Sandberg U, Mun EP (2007) The little book of quieter pavements. FHWA, Austin, Texas

A. A. Bawono, *Engineered Cementitious Composites for Electrified Roadway in Megacities*, Springer Theses, https://doi.org/10.1007/978-3-030-88542-7

Table E.1 Pavement surface texture test method

Test method/associated standard	Description	Applicability
Sand patch method (SPM); ASTM E965, ISO 10844	This volumetric-based test method provides the mean depth of road surface macrotexture. The operator spreads a known volume of glass beads in a circle onto the surface and determines the diameter and mean texture depth (MTD)	• Simple and inexpensive • Localized method • Accuracy 2%
Outflow (OF) meter; ASTM WK 364	Volumetric test method that provides a measure of the escape time for water beneath a moving tire. The operator measures the rate of gravity-controlled outflow from a cylinder placed on a road surface	• Simple methods and relatively inexpensive equipment • Localized measurement • Accuracy 0.01 s
Circular texture (CT) meter; ASTM E 2157	Provides a mean profile depth of the road surface macrotexture. The equipment measures a 284 mm circular profile of the road surface at intervals of 0.87 mm	• Measures same diameter as DFT • Localized measurement • Texture measured in 2 directions • Accuracy 0.03 mm
Texture depth gauge (TDG); ASTM T 261	Provides an average depth of PCC grooves or tines. The gauge is inserted into 10 grooves to measure their depths	• Simple methods and inexpensive equipment • Not a measure of defined texture
British Pendulum Tester (BPT); ASTM E303	Provides an indirect measure of relative microtexture. The testing device measures drag on a rubber footed pendulum swung across the road surface	• Methodology is critical • Can be done in laboratory • Accuracy 1.2 BPN
Electro-optic (laser light sectioning, ultrasonic, stylus) method (EOM); ASTM E 1845, ISO 13473-1, ISO 13473-2, ISO 13473-3	Provides a profile of the road surface macrotexture. This equipment uses an optical distance measuring sensor to collect surface elevation data at intervals of 0.25 mm or less	• Some equipment collects at high speeds • Correlates well with MTD • Continuous measurement possible • One direction only • Accuracy 0.15 mm

Table E.2 Summary of texturing methods

Application	Mix/texture type	Description
New AC or AC overlay	Dense fine-graded HMA	Dense-graded HMA is a dense, continuously graded mixture of coarse and fine aggregates, mineral filler, and asphalt cement (5–6%). Dense-graded HMA can be modified with polymers or crumb rubber, and may include recycled materials. Nominal MAS range from 9.5 to 19.0 mm
	Dense coarse-graded HMA	Coarse HMA mixes have gradations that pass below the MDL at the No. 8 sieve (2.36-mm)
	Gap-graded HMA or stone matrix asphalt (SMA)	SMA is a gap-graded mixture of course aggregate (typically, 10–15 mm), filler, fibers and polymer-modified asphalt. Its primary advantage is resistance to deformation, but its relatively coarse surface yields good frictional characteristics
	Open-graded friction course (OGFC)	OGFC is an open-graded mixture of mostly coarse aggregate, mineral filler, and asphalt cement (3–6%). Open-graded HMA can be modified with polymers, fibers, and/or crumb rubber
Friction restoration of existing AC pavement	Chip seal	Thin surface treatment containing single-sized, high-quality, angular aggregates (9.5–15 mm), spread over and rolled into a liquid asphalt or asphalt emulsion binder. Aggregates are sometimes pre-coated with asphalt emulsion prior to spreading. Completed surface is somewhat coarse, yielding good frictional characteristics
	Slurry seal	Slurry mixtures of fine aggregate, mineral filler, and asphalt emulsion. They are similar to micro-surfacing, without interlocking aggregates. Polymers are not always used in the emulsion. Their surface is typically gritty
	Micro-surfacing (polymer-modified slurry seal)	A slurry mixture containing high-quality crushed, dense-graded aggregate, mineral filler, and polymer-modified asphalt emulsion. It is placed over a tack coat and is capable of being spread in variable thickness layers for rut-filling, correction courses, and wearing course applications
	HMA overlay	See HMA surface mixes above
	Ultra-thin polymer-modified asphalt (e.g., NovaChip)	Thin gap-graded asphalt surfaces placed using specialized equipment immediately over a thick polymer-modified asphalt emulsion membrane. Following slight compaction the surface provides a semi-porous texture
	Epoxied synthetic treatment (e.g., Italgrip)	A very thin surface treatment consisting of a two-part polymer resin placed on an existing pavement and covered with a man-made aggregate of re-worked steel slag (3–4 mm). The surface is designed to substantially improve the frictional characteristics of pavements

(continued)

Table E.2 (continued)

Application	Mix/texture type	Description
Retexturing of existing AC pavement	Micro-milling	Milling equipment, consisting of a self-propelled machine with carbide teeth mounted on a rotating drum, typically removes 19–32 mm from the asphalt surface. The spacing of cuts is approximately 5 mm versus 6 mm cut of conventional cold-milling machines. The resulting surface has a fine, smooth pattern that gives a smoother ride
New PCC or PCC overlay	Broom drag (longitudinal or transverse)	A long-bristled broom is mechanically or manually dragged over the concrete surface in either the longitudinal or transverse direction. Texture properties are controlled by adjusting the broom angle, bristle properties (length, strength, density), and delay behind the paver. Uniform striations approximately 1.5–3.0 mm deep are produced by this method
	Artificial turf drag (longitudinal)	An inverted section of artificial turf is dragged longitudinally over a concrete surface following placement. Texture properties are controlled by raising/lowering the support boom, adding weight to the turf, and delaying application to allow surface hardening. This method produces uniform 1.5–3.0 mm deep surface striations
	Burlap drag (longitudinal)	One or two layers of moistened coarse burlap sheeting are dragged over the concrete surface following placement. Texture properties are controlled by raising/lowering the support boom and adjusting the delay following concrete placement
	Longitudinal tine	A mechanical assembly drags a wire comb of tines (~127 mm long and 3 m wide) behind the paver (and usually following a burlap or turf drag). Texture properties are controlled by the tine angle, tine length, tine spacing, and delay for surface curing. Grooves from 3–6 mm deep and 3 mm wide are produced by this method, typically spaced at 19 mm
	Transverse tine	Accomplished using methods similar to longitudinal tining, however, the mechanical assembly drags the wire comb perpendicular to the paving direction. Variations include skewing the tines 9°–14° from perpendicular and using random or uniform tine spacing from 12 to 38 mm
	Diamond grinding (longitudinal)	A self-propelled grinding machine with a grinding head of gang-mounted diamond sawing blades removes 3–19 mm of cured concrete surface, leaving a corduroy type surface. Blades are typically 2–4 mm wide and spaced 4.5–6 mm apart, leaving 2–4 mm high ridges. This method is most commonly used to restore surface characteristics of existing pavements
	Porous PCC	Gap-graded, small-diameter aggregate are combined with cement, polymers, and water to form a drainable surface layer (typically 200 mm thick). That surface layer is bonded to the underlying wet or dry dense concrete layer

(continued)

Table E.2 (continued)

Application	Mix/texture type	Description
	Exposed aggregate PCC	A set retarder is applied to the wet concrete surface and the surface is protected for curing. After 12–24 h, the unset mortar is removed to a depth of 1–2 mm using a power broom
Friction restoration of existing PCC pavement	HMA overlay	See HMA surface mixes above
	TyreGrip	Tyregrip® is a resurfacing system that consists of a two-part epoxy resin top dressed with calcined bauxite aggregate. The aggregate is fractured 100% and is approximately No. 10 (2 mm) size material
	Italgrip	
	White topping, thin white topping (TWT) and ultra white topping (UTW)	White topping is a PCC layer constructed atop an existing HMA pavement structure [2] Conventional whitetopping—a concrete overlay of 200 mm or more, designed and constructed without consideration of a bond between the concrete and underlying HMA TWT—an overlay of greater than 100 mm and less than 200 mm in thickness UTW—with a thickness equal to or less than 100 mm (4 in.), this overlay requires a bond to the underlying HMA to perform well
Retexturing of existing PCC pavement	Diamond grinding (longitudinal)	See diamond grinding above
	Longitudinal diamond grooving	A self-propelled grooving machine saws longitudinal groove in the road surface about 3–6 mm deep and spaced 13–38 mm apart. This method adds macrotexture for drainage but relies on the original surface for microtexture
	Transverse diamond grooving	Completed in a manner similar to longitudinal diamond grooving, except the grooves are sawn transverse to the travel direction
	Shot abrading	An automated machine hurls recycled round steel abrasive material at the pavement surface, abrading the surface and/or removing the mortar and sand particles surrounding the coarse aggregate to a depth of up to 6 mm. Texture properties are controlled by adjusting the steel abrasive material velocity and approach angle and by modifying the forward equipment speed

About the Author

Ali Aryo Bawono is currently a Postdoctoral Fellow at TUMCREATE Limited in Singapore. Prior to his current role at TUMCREATE, Bawono was a Research Associate at the Technical University of Munich, Germany. At the same time, he did joint research with the Nanyang Technological University of Singapore. He has also spent more than five years in the infrastructure industry, with four years' experience as a Senior Civil Engineer in the Ministry of Public Works and Public Housing of the Republic of Indonesia and a year as a Chief Engineer for a toll-highway project at PT Jaya Konstruksi, one of the biggest construction company in Indonesia.

This book is the result of Bawono's Doctoral Thesis in Civil Engineering in the Department of Civil, Geo, and Environmental Engineering at the Technical University of Munich, Germany. Bawono has also received a Bachelor of Science in Civil Engineering from The Institute of Technology Bandung, Indonesia, and a Master of Science in Transportation System from the Technical University of Munich, Germany.

Bawono has published over ten top-tier journal articles and peer-reviewed conference papers, given invited talks and seminars at several workshops and institutions such as TUMCREATE in Singapore, Transportation Research Board in Washington D.C., International Conference on SEMC by the University of Cape Town in South Africa, University of Tarumanegara, University of Trisakti, and the National University of Surabaya in Indonesia. He holds patent and technology disclosures. He was the finalist for the best idea in the transportation infrastructure theme competition awarded by the Mayor City of Bandung, Indonesia.

His research interest includes conventional and electrified roadway, smart infrastructure, Engineered Cementitious Composites material, self-cleaning (photocatalysis) material, Building Information Modelling, Multi-Criteria Design Analysis, Machine Learning and Deep Learning Method.

A. A. Bawono, *Engineered Cementitious Composites for Electrified Roadway in Megacities*, Springer Theses, https://doi.org/10.1007/978-3-030-88542-7

Printed by Printforce, the Netherlands